Science and the State

Was it coincidence that the modern state and modern science arose at the same time? This overview of the relations of science and state from the Scientific Revolution to World War II explores this issue, synthesising a range of approaches from history and political theory. John Gascoigne argues the case for an ongoing mutual dependence of the state and science in ways which have promoted the consolidation of both. Drawing on a wide body of scholarship, he shows how the changing functions of the state have brought a wider engagement with science, while the possibilities that science makes available have increased the authority of the state along with its prowess in war. At the end of World War II the alliance between science and state was securely established and, Gascoigne argues, is still firmly embodied in the post-war world.

Emeritus Professor John Gascoigne taught history at the University of New South Wales from 1980 until 2016. This is the sixth of his books with Cambridge University Press, which include *Encountering the Pacific in the Age of the Enlightenment*, which won the NSW Premier's General History Prize in 2014, and *Science in the Service of Empire: Joseph Banks, the British State and the Uses of Science in the Age of Revolution*.

New Approaches to the History of Science and Medicine

This dynamic new series publishes concise but authoritative surveys on the key themes and problems in the history of science and medicine. Books in the series are written by established scholars at a level and length accessible to students and general readers, introducing and engaging major questions of historical analysis and debate.

Science and the State

From the Scientific Revolution to World War II

John Gascoigne

University of New South Wales, Sydney

 **CAMBRIDGE** UNIVERSITY PRESS

CAMBRIDGE
UNIVERSITY PRESS

University Printing House, Cambridge CB2 8BS, United Kingdom

One Liberty Plaza, 20th Floor, New York, NY 10006, USA

477 Williamstown Road, Port Melbourne, VIC 3207, Australia

314–321, 3rd Floor, Plot 3, Splendor Forum, Jasola District Centre, New Delhi – 110025, India

79 Anson Road, #06–04/06, Singapore 079906

Cambridge University Press is part of the University of Cambridge.

It furthers the University's mission by disseminating knowledge in the pursuit of education, learning and research at the highest international levels of excellence.

www.cambridge.org
Information on this title: www.cambridge.org/9781107155671
DOI: 10.1017/9781316659120

First published 2019

Printed in the United Kingdom by TJ International Ltd. Padstow Cornwall

A catalogue record for this publication is available from the British Library.

Library of Congress Cataloging-in-Publication Data
Names: Gascoigne, John, 1951– author.
Title: Science and the state : from the scientific revolution to World War II / John Gascoigne.
Description: Cambridge, United Kingdom ; New York, NY : Cambridge University Press, 2019. | Includes bibliographical references.
Identifiers: LCCN 2018042040| ISBN 9781107155671 (hardback) | ISBN 9781316609385 (paperback)
Subjects: LCSH: Science and state – History.
Classification: LCC Q125.2 .G385 2019 | DDC 509/.03–dc23
LC record available at https://lccn.loc.gov/2018042040

ISBN 978-1-107-15567-1 Hardback
ISBN 978-1-316-60938-5 Paperback

For my brother, Robert

Contents

Illustrations

Abbreviations

CNRS	Centre National de la Recherche Scientifique (the National Centre for Scientific Research)
DSIR	Department of Scientific and Industrial Research
KW	Kaiser Wilhelm
MITI	Ministry for International Trade and Industry
NDRC	National Defence Research Council
NIH	National Institutes of Health
NSF	National Science Foundation
ONR	Office of Naval Research
OSRD	Office of Strategic Research and Development

Preface

The rise of science has been linked with a number of major trends in the history of the West: the rise of capitalism, the growth of Protestantism or the impact of exploration and contact with new lands. The object of this book is to examine another possible linkage: the extent to which the formation and consolidation of science was associated with that distinctive feature of Western history, the rise of the state. This is a subject that has received scholarly attention, but which has not been accorded the full-length examination over time that it merits. For the aim of this work is to examine the relationship between the state and science from the age of the Scientific Revolution until the end of World War II. What the book seeks to show is how reciprocal was the linkage between the state and science, with science in some respects strengthening the ideological and institutional reach of the state and, on the other, the ways in which state structures moulded the shape of science through its patronage of particular strands of scientific endeavour.

The work begins with some consideration of the character of both science and the state. Of particular interest here are the characteristics which promoted the formation of the modern state and its entrenchment as a global phenomenon (Chapter 1). This is followed by an examination of the early form of the state in the period of the Renaissance and the extent to which royal and princely patronage offered a basis for partnership between science and the state (Chapter 2). The Renaissance state is then contrasted with the absolutist state that followed it and that provided securer institutional foundations for building an alliance between the state and science (Chapter 3). On the other hand, nonabsolutist states are also shown to have developed different but nonetheless significant forms of linkage between their forms of government and the state (Chapter 4). The old-regime world of the absolutist state was shattered by the age of revolutions and, above all, by the French Revolution, which led to new state structures and ideologies that, in many cases, further entrenched science in the apparatus of government (Chapter 5). By the time of the period from the mid-nineteenth century to World War I, the state was widening its functions in ways that promoted more recourse by government to scientific advice, with the beginnings of systems of social welfare and public health. This was also the period of 'high imperialism' which meant a greater attention to the uses of science as a device for extending and consolidating imperial rule. (Chapter 6). The period from 1914 to 1945 is, of course, dominated by war, which did much to consummate fully the partnership between science and the state. It was also a period when science came under the sway of totalitarian governments, an experience which underlined the adaptability of scientists to various regimes (Chapter 7). By 1945, both science and the state had spread to much of the globe, taking different forms as it did so. Global expansion also raised issues about the tension between the national and the international dimensions of science (Chapter 8). The integration of science and government, which was brought to pass by the war, makes 1945 a natural ending point for this survey. To reinforce this argument an Epilogue considers some of the major developments that occurred around the globe in the decades immediately after World War II to illustrate how 1945 marked a watershed. Thereafter, there were, of course, significant developments, but they occurred within the context of a consensus that the state had a major role to play as a patron of science. To make this work of manageable size, the method generally followed is to provide case studies of key countries in comparison with each other in the context of the general themes which I have outlined. The result is the omission of other countries that might have been included, but perhaps will be the subject of future studies by others to advance further our understanding of the relations between science and the state.

As a work of synthesis this book is much indebted to many scholars who have written on topics linked to its major themes. My thanks go to Lucy Rhymer at Cambridge University Press for her interest in this project. My thanks, too, to my copy editor, Matthew La Fontaine, for his close attention to my text and to Arnia Van Vuuren for producing a comprehensive index. A one-month fellowship at the Huntington Library, Los Angeles, enabled me to begin work on this book in its pleasant surroundings. Thanks to my family, my children, Robert and Catherine, and my wife, Kate, for their support during the ups and downs of the period when this book was being researched and written.

Chronology

1543	publication of Nicolaus Copernicus' *About the revolutions of the heavenly spheres* and Andreas Vesalius's *About the structure of the human body*
1610	Galileo appointed Mathematician-Philosopher at the court of the Medicis
1626	establishment of the French Royal Garden
1627	publication of Francis Bacon's *New Atlantis*
1633	papal condemnation of Galileo for advancing the Copernican theory
1651	publication of Thomas Hobbes' *Leviathan*
1660	foundation of the Royal Society of London
1666	foundation of the Academy of Sciences of Paris
1687	publication of Isaac Newton's *Mathematical principles of natural philosophy*
1714	British Board of Longitude established
1724/5	foundation of the Saint Petersburg Academy of Science
1736 to 1737, 1735 to 1743 respectively:	Pierre Maupertuis' voyage to Lapland, Charles-Marie de La Condamine's voyage to Peru – both sponsored by Academy of Sciences of Paris in order to determine shape of the earth
1743	foundation of the American Philosophical Society
1744	refounding by Frederick the Great of the Berlin Royal Academy of Sciences
1773	the Royal Gardens at Kew transformed from a royal pleasure garden to a botanical research institute
1774	Spain establishes a royal garden
1793	closure of the Academy of Sciences by the French revolutionary government
1793	foundation of the National Museum of Natural History (replaces the Royal Garden)

1794	foundation of the École Polytechnique
1795	foundation of the French National Institute
1795	foundation of the French Bureau des longitudes
1863	foundation of US National Academy of Sciences
1883	foundation of the British Museum of Natural History
1887	foundation of the German Imperial Physical Technical Institute
1888	foundation of the Paris Pasteur Institute for biological research
1891	foundation of the Berlin Robert Koch Institute for disease control
1900	foundation of the British National Physical Laboratory
1901	foundation of the US Bureau of Standards
1901	foundation of the Rockefeller Institute for Medical Research
1911	foundation of the Kaiser Wilhelm Society for the Promotion of Research
1916	foundation of the US National Research Council
1916	foundation of the British Department of Scientific and Industrial Research (DSIR)
1929	Stalin's 'Great Break'; beginning of purge of the Soviet Academy of Sciences
1939	foundation of French Centre national de la recherche scientifique (CNRS) (the National Centre for Scientific Research)
1940	foundation of the US National Research Defence Committee
1941	foundation of the US Office of Scientific Research and Development
6 Aug 1945	detonation of the atomic bomb over Hiroshima
1945	publication of *Science – the endless frontier*
1946	foundation of US Office of Naval Research
1948	renaming of the Kaiser Wilhelm Society, the Max Planck Institute
1949	detonation of Soviet atomic bomb
1950	formation of the US National Science Foundation (NSF)

1 Introduction

Science, the State and Their Mutual Dependence

Modern science and the modern state emerged at much the same time in early modern Europe, and both institutions were consolidated further in the centuries which followed – particularly so in the nineteenth and twentieth centuries in response to the imperatives of industrialisation and war. Was this coincidence? It is the argument of this book that it was not, that the growth of science and the state were linked, and that both drew on each other in establishing and augmenting their sway. To convey an overview of the major themes that such a survey of the relations between science and the state entails, we begin by asking what, in broad terms, were some of the primary ways in which the state and science interacted?

Firstly, what has science sought from the state? Most obviously patronage in the form of material benefits, but also protection and the conferral of prestige. Such state patronage was to extend to royal charters of scientific associations, which brought with them a measure of control over the conduct of science.[1] This recognition by the state was particularly important in the early stages of the modern scientific movement, when other possible patrons, notably the church and the aristocracy, showed only a limited interest in science. Looking to the state, which more and more was acquiring an impersonal identity that transcended the role of particular rulers, meant some distancing of science from the dynastic upheavals to which individual monarchs were heir.[2] In its early stages, science was a culturally marginal activity with an ambivalent relationship with the two chief intellectual authority structures of Western Europe, the Bible and the classical heritage. The state to some extent provided a counterweight to such engrained cultural

[1] Hahn, 'The age of academies', pp. 6–7
[2] Porter, 'The scientific revolution', pp. 290–316

1

traditions – what acknowledgement was accorded by the state to the scientific movement gave its practitioners a self-confidence and sense of worth that enabled science to take root and to grow in size and stature.

But, with state munificence came forms of control, something that was much less palatable to the scientists (to use a word not coined by William Whewell until 1833),[3] and was the source of an ongoing tension between the state and the scientific community. Different forms of state sought to deal with this tension in different ways – hence a theme of the book is the way in which, within the different forms of state that developed from the sixteenth century onwards, science also developed different institutional forms. Another source of tension between scientists and the state is dealing with both the national and international aspects of science. Science by its nature is international in character, and scientists seek the stimulation of contact with their fellow specialists whatever their nationality. Louis Pasteur once famously said that 'scientists have a country; science has none'.[4] Yet, as subjects or citizens and, in many cases, as beneficiaries of state largesse, scientists are subject to individual states which, in some times and places, have discouraged or forbidden international scientific contact as detracting from national goals. Like all relationships, then, that between science and the state has had and continues to have points of tension, but these have been outweighed by the need to maintain a mutually advantageous alliance.

What has the state asked of science? First and foremost, the state has sought practical advantage from science in the promotion of utilitarian or military goals. The fact that the intersection between science and such practical ends was not well developed until the nineteenth century meant that state support for science was accordingly rather muted until then. But, along with the hope that the promotion of theoretical understanding of nature might lead to ways of commanding nature for human or, at least, governmental advantage, the state sought other advantages from science. Just as the state could confer some measure of prestige on science, so too could science bring to the state reflected glory by association with scientific findings. This was particularly important in the early stages of the state's development, when its means of establishing its authority both internally and externally were weak, necessitating a reliance on cultural capital to increase its sway. In this sense, patronage of science played a similar role to patronage of literature or art in augmenting the prestige of the state. More fundamentally, the state has looked to science to reinforce its own claims to authority. The state developed in the context of early modern Europe, when the traditional structures of political authority

[3] Reidy, *Tides of history*, p. 242   [4] Von Gizycki, 'Centre and periphery', p. 477

based around the church had been undermined by the forces unleashed by the Reformation.[5] Scientific reasoning provided another basis for politics and, in particular, for a state which claimed to be characterised by rationality and order, just as science was. As the concept of the state more and more transcended the personality of individual rulers, it developed an abstract character that meshed well with scientific speculation on the abstract body of laws that controlled the workings of nature.

The emerging state also shared with science a preoccupation with knowledge, which was made manifest by the increasingly bureaucratic character of government. The growing mounds of paper through which government was conducted represented the attempt to centralise and systemise knowledge, the better to bring the larger society under the government's sway. The role of science in helping to shore up the authority of the state was to be taken further as the state developed. Post-revolutionary regimes, particularly, looked to science as a form of ideology to justify their break with the past. Science and the new centralising regime were viewed as partners in the modernisation of society against the traditional forces of ignorance and localised privilege.[6] Such a science-based ideology was to be employed, too, to justify states' imperial expansion and the imposition of forms of colonial rule that were claimed to bring the benefits of modernisation to non-Western states.[7]

Both at home and abroad, the goals of the modernising state promoted the emergence of the social sciences as a means of imposing greater social order and economic advance.[8] The emerging social sciences were given a scientific and mathematical edge through the use of what became known as 'statistics'. The origins of the term derive from eighteenth-century German cameralism, a school of statesmanship aiming at developing a science of administration. The association of the word 'statistics' with the state is evident in the way it derived from the German 'staat', with the term 'statistics' being first used by the Göttingen professor, Gottfried Achenwall, in 1749.[9] One characteristic product of cameralism was a science of forestry that aimed to promote both increased predictability and order together with productivity.[10] As such methods gained ground with the expansion of the state's power and growing commitment to the imposition of Enlightenment values, the result was increasingly to be what David Resnik calls a '"scientisation" of politics'.[11]

---

[5] Shapin, *The scientific revolution*, p. 123   [6] McClellan, *Science reorganised*, p. 26
[7] Sivasundaram, 'Sciences and the global', p. 154
[8] Godlewska, *Geography unbound*, p. 193   [9] Porter, *The rise*, p. 23
[10] Scott, *Seeing like a state*, p. 14   [11] David Resnik, *Playing politics*, p. 10

Defining Science

In addressing the wide-ranging issues entailed in discussing the relationship between science and the state, one necessarily begins with some definitions. The English word 'science' is confined to knowledge of the natural world, in contrast to its equivalent in most Western European languages, which have a much broader meaning encompassing all forms of disciplined learning including fields such as history. The character of science as a way of interrogating the natural world is usefully summarised by Simon Schaffer as being an 'organised methodical investigation of nature's capacities'.[12] As Hilary and Stephen Rose point out, however, the term 'science' is used in different senses.[13] The first three nuances of the word they summarise as:

(i) the pursuit of natural laws
(ii) the application of certain rules of procedure and inquiry
(iii) the social institutions with within which the activity is carried out.

Of these meanings, the one most directly pertinent to this study is the third one, linking science with its institutions – institutions which, in their turn, were generally to involve some sort of liaison or, at least, accommodation with the agencies of the state. The Roses also give two more meanings of the word 'science' that underline the uncertainties of the boundary line between 'science' and 'technology' – a term not coined until 1831 (by Jacob Bigelow):[14]

(iv) as including the whole field of research and development, that is, both science and technology
(v) as excluding the technological development of science, embracing instead only pure science.

Of these, number four best encompasses the way in which the word 'science' is used in this work. The separation between science and technology was more evident before the nineteenth century, when technological innovation became more closely allied with scientific advance.[15] In all periods, however, there was some overlap between the methods and goals of both, and this interaction became progressively more pronounced. It is true to say, as Jagdish Sinha writes, that 'generally speaking, the understanding of nature is science and the means and methods of manipulating that understanding for specific purposes is technology'.[16] What unites both, however, is a common assumption that nature is law-abiding or, at least, orderly and that this characteristic can serve as a basis for human command over nature. In this book, then, the domain of

---

[12] Schaffer, 'What is science?', p. 27   [13] Rose, *Science and society*, p. 2
[14] Smith, *American science policy*, p. 23   [15] Smith, *American science policy*, p. 29
[16] Sinha, *Science, war and imperialism*, p. 3

science as an ally of the state will be viewed in the wider sense of incorporating both the understanding and manipulation of nature in ways that paralleled the increasingly centralised and ordered character of the state.

Defining the State

The state was pithily defined by the great sociologist, Max Weber, as the entity exercising 'the monopoly of the legitimate use of physical force within a given territory' – a summation which underlines the central feature of the state, that it is a sovereign entity based on a territorial form of political organisation.[17] The state is not the same as 'government', the apparatus by which the state achieves its ends, for government can exist in forms of rule other than a 'state', such as feudalism or an empire. The emergence of states brought with it not only increasing centralisation within states, but also a changing relationship between states as a new international order developed to deal with the rivalry between states.[18]

Understanding the different forms the state has taken is, however, a more involved task that requires surveying in summary fashion the development of that increasingly important institution. The attempt to impose a unitary rule over Western Europe in the Middle Ages failed as the two putative heads, the Holy Roman Emperor and the pope, vied with each other for dominance. The ensuing vacuum left room for the growth of individual states, which were to be in a constant state of rivalry with one another. Such rivalry generated the need for greater military forces, which, in turn, necessitated higher taxation and the increasing bureaucratisation of the state this entailed. This process is most clearly evident in the rise to power of one of the most successful absolutist states, Prussia, which developed a science of politics, *kameralwissenschaft*, to promote the efficiency of its military-fiscal bureaucratic machine.[19] The nexus between war and the development of the state was captured succinctly by Charles Tilly in his telling remark: 'War made the state, and the state made war.'[20]

The needs of war, then, tended to push the development of the state towards an absolutist form with power concentrated at the centre. There were exceptions, however, as some states retained the originally feudal forms of parliamentary rule. The most notable exception in this regard was England, which, as an island with a natural moat, did not have to

[17] Nelson, *The making of the modern state*, p. 7
[18] Vincent, *Theories of the state*, pp. 31, 225   [19] Poggi, 'Modern state', p. 342
[20] Tilly, 'Reflections', p. 43

maintain a large standing army, though the costs of a growing navy led to an increasing bureaucratisation of the eighteenth-century state.[21] The English common law tradition was also less of a natural ally of absolutism than the Roman law based around the power of the emperor that predominated on the Continent.[22] Such a divergence was of particular importance since the ideal of an impersonal state was closely linked to the state as a legal structure. This was given greater force with the recovery of Roman law in the late Middle Ages and Renaissance acting as a stimulus to the development of Continental absolutism.[23]

Conceptualising the development of the state posed particular problems for contemporaries. For centuries, people had thought in terms of the person of the ruler rather than the state, with the ruler's territories a dynastic possession that changed hands with the fortunes of individuals. Consistent with such a view, ambassadors were regarded as personal representatives of the prince.[24] As the state brought more and more of society under its sway, there was a gradual shift from viewing the sovereign as an individual to regarding the ruler as the embodiment of the state. The next and most important shift that marked the emergence of the modern state was conceiving of the state as an impersonal entity that continued on as individual monarchs and their systems of government came and went. As such, the state could serve as the guardian of a constitutional order that increasingly claimed to be based on rational principles that transcended individual preferences.[25] Like science, politics came to be based more and more on abstractions.[26]

The growth of the state bureaucracies meant that the conduct of state power came more and more to be characterised by what Max Weber termed 'the exercise of control on the basis of knowledge', making scientific knowledge increasingly relevant to the concerns of the state. As these bureaucratic machines loomed ever larger, they further promoted the idea of the state as an impersonal entity that could continue to operate in the absence of the individual ruler.[27] A further and later development, once the apparatus and legal authority of the state had been established, was the emergence of the 'nation state', particularly as nationalist sentiment grew in the wake of the French Revolution. By 'nation' was meant a group of people sharing a common culture and, generally, a common language. A 'nation state' combined such bonds of unity with the political standing and authority of the state.[28]

---

[21] Brewer, *Sinews of war*   [22] Dyson, *The state tradition*, p. 42
[23] Nelson, *The making*, pp. 26–7   [24] Shennan, *The origins*, p. 34
[25] Vincent, *Theories of the state*, pp. 50, 79   [26] Tivey, 'Introduction', p. 3
[27] Soll, *The information master*, p. 14   [28] Dyson, *The state tradition*, p. 129

Thomas Hobbes, the State and Science

Perhaps the most influential theorist of the modern state was Thomas Hobbes (1588–1679). In his *Leviathan* (1651), he was the first to define the 'state' in impersonal terms that clearly distinguished it from the person of the ruler.[29] Hence, allegiance was required to an abstract concept rather than an individual or a community.[30] Accordingly, he wrote in the preface to the Latin edition of the *Leviathan*: 'This great Leviathan, which is called the State, is a work of art [i.e. man-made]; it is an artificial man made for the protection and salvation of the natural man, to whom it is superior in grandeur and power.'[31] Indeed, in Hobbes's understanding of the state, it was impersonal in two senses: firstly since it transcended individual rulers who might govern for a time, and secondly because it was distinct from the community over which the Leviathan ruled.[32] He was prompted to make this conceptual shift to an impersonal state by the dissension into which England had descended during the civil wars of the mid-seventeenth century. These he saw as the outcome of a political system that had lost its traditional political moorings and required a foundation based on the sort of reasoning which Hobbes so relished in mathematics.[33] The desire for a political system with a clear structure organised around a central principle made Hobbes particularly susceptible to absolutism, even though his native England was to retain a parliamentary system which sat uneasily in Hobbes's system.[34]

Political and scientific theorising were, to Hobbes, two sides of one coin sharing a common materialistic method. It was this method which made political theorising, as Hobbes wished to practise it, a new discipline. Indeed, he saw himself as the founder of a true science of politics, writing that 'Natural Philosophy is therefore but young; but Civil Philosophy yet much younger, as being no older . . . than my own book *De Cive* [*About the state*].'[35] Both natural philosophy and the science of politics depended on the study of separate particles in motion, either as material bodies in the case of science, or individual human beings with their emotions constituting a form of motion. As he wrote in his major treatise on science (or as the seventeenth century termed it, 'natural philosophy') *De Corpore* (*About the body*): 'After physics we must come to *moral philosophy*; in which we are to consider the motions of the mind.'[36] Indeed, in *Leviathan* there was an analogy between Hobbes's method and that of

[29] Nelson, *The making*, p. 69   [30] Cudworth, *The modern state*, p. 23
[31] Nelson, *The making*, pp. 188–9   [32] Skinner, 'The state', p. 112
[33] Cornelia Navari, 'The origins', p. 20   [34] Skinner, 'The state', p. 116
[35] Macpherson, 'Introduction', p. 10   [36] Finn, *Thomas Hobbes*, p. 48

1.1 Frontispiece to Thomas Hobbes's *Leviathan*

Galileo (whom Hobbes met in 1634), with both resolving their subject into individual parts moved by forces.[37]

This nexus between Hobbes's political and natural philosophy has been emphasised by Stephen Shapin and Simon Schaffer in their classic study, *Leviathan and the Air Pump*. Here they draw out the way in which experimental data, of the sort derived by Robert Boyle from the air pump, did not meet Hobbes's canons of deductive certainty that he sought in both politics and nature. Allowing experimental evidence was, for Hobbes, to weaken the geometrical-like structure that he viewed as essential to the maintenance of a secure political order based around an impersonal, sovereign, law-enforcing state.[38] For, in the face of the political discord which had led to civil war, it was Hobbes's ambition to construct 'a science . . . built upon sure and clear principles' and 'to reduce [politics] to the rules and infallibility of reason'.[39]

Hobbes's work, then, reflects clearly the emergence of the new understanding of the state as abstract and impersonal – something that, to Hobbes's way of thinking, was linked to the emergence of a new natural philosophy. Of course, across Europe, different states were at different stages in the continuum from individual princely rule to defining themselves in impersonal terms. It has been argued by the prominent theorist of the state, Charles Tilly, however, that there was greater convergence in the forms of the state in the nineteenth and twentieth centuries.[40] This perhaps reflects the fact that by this later period, states could more uniformly draw on a greater body of knowledge of the composition of their territories and the character of their population made possible by the application of scientific methods to the task of state-building. With greater knowledge came an expansion of the state's functions in the nineteenth and twentieth centuries to encompass the lives of their citizens or subjects to an extent hitherto unparalleled.[41] By contrast, when the modern state came into being, its defining task was more narrowly defined as the administration of justice: hence Hobbes saw law as 'the will and Appetite of the State'.[42]

State and Science in a Global Perspective

Gradually, then, the state became the dominant political form within Europe, and the interplay between such states formed the substance of European international relations. European imperial power imposed

---

[37] Finn, *Thomas Hobbes*, p. 12   [38] Shapin and Schaffer, *Leviathan*, pp. 29, 333, 344
[39] Macpherson, 'Introduction', p. 10   [40] Tilly, 'Reflections', p. 34
[41] Cudworth, *The modern state*, p. 34   [42] Hobbes, *Leviathan*, p. 694

state forms on much of the rest of the world, with the colonial state and science often working together to promote forms of modernisation which corresponded to the interests of colonial powers. Yet, in the spheres of both science and state formation, indigenous culture often gave the European-derived forms a new character. There were also parts of the world such as Japan or Thailand, unmarked by imperialism, which developed state forms from indigenous roots together with judicious borrowings from the West. One of the great ironies of imperialism is that the construction of new imperial states often provided the institutional platform for the development of movements for independence.[43] However, as in Africa, imperial powers sometimes exported states rather than nation states. The result was a state with bureaucratic and legal authority, but with an unstable mixture of different cultures and tribal identities, which has led to internal unrest.[44]

The pressures to maintain some form of state are strong, as statehood has become an accepted preliminary to joining an international order that has been based on reconciling the interests of different states.[45] The establishment of the state by imperial regimes was, too, linked to the modernising impulses of science. Yet such imperially linked structures have been turned against imperialists, for not only the state but science, too, has been co-opted as a force for liberation from Western domination. Both science and the state were to prove willing and able to serve different masters as they more and more encompassed the globe.

---

[43] Anderson, *Imagined communities*, p. 127; Tinkler, 'The national state in Asia', p. 119
[44] Hughes, 'The nation-state', p. 142   [45] Tilly, 'Western state-making', p. 637

2   The Renaissance Monarchy

The Early Scientific Movement and its Patrons

Many streams converged to form what became the Scientific Revolution. This is generally defined as beginning with the publication of Copernicus's *Revolutions of the Heavenly Spheres* and Vesalius's *The Structure of the Human Body* in 1543 and coming to its culmination with the publication of Newton's *Principia* in 1687. Over the centuries diverse cultures had contributed forms of learning which were amalgamated in the early scientific movement: Sumerian number systems, Egyptian astronomy, Greek study of many aspects of nature, Indian systems of numerals, Arabic learning together with its transmission of classical texts and the translation into Latin by Jewish scholars of many of these key texts. The increasing movement between cultures was accelerated in the early modern period by growing contact between hitherto largely isolated parts of the globe. This was heightened by the series of voyages begun by Vasco da Gama's rounding of Africa to reach India from Europe in 1498, and Christopher Columbus's first epochal voyage to America in 1492, which brought a whole new continent into European consciousness. What were, from a European perspective, discoveries of new parts of the globe brought with it a mentality more willing to explore the natural world in hitherto unparalleled ways, which went beyond the frontiers of learning that had been established by the Ancients.[1]

By its nature, the scientific movement, as a departure from the established ways of thinking, did not fit naturally into the existing structures of learning. Its most obvious home was the universities, which had cultivated some forms of natural philosophy throughout the Middle Ages. But the universities did not prove an altogether congenial home to the Scientific Revolution, even though most of the major figures linked to the early scientific movement had some association with the universities,

---

[1] Gascoigne, 'Crossing'

either as students or professors.[2] The universities were too shackled to traditional forms of philosophy based on Aristotle's work to absorb readily the insights of the Scientific Revolution. For the scholastic philosophy of the universities based on logical deduction from authoritative texts left only a limited role for empirical evidence. It also separated natural philosophy from mathematics, in contrast to the impulse of the Scientific Revolution to combine the two.

Another possible patron of the Scientific Revolution was the Catholic Church and, indeed, many major figures of the early scientific movement (among them Nicolaus Copernicus and Pierre Gassendi) held clerical posts. Their scientific interests were, however, an avocation rather than being prescribed by the nature of their employment leaving limited room for anything approaching a secure scientific career. The Renaissance papacy, which came to resemble other European monarchies, extended its patronage from the arts to some branches of science: one of the holders of the much-prized post of papal physician, for example, was Marcello Malpighi, discoverer of the capillaries.[3] Another instance is the publication of the first volume of *Ornithology* (1599) by the pioneering naturalist, Ulisse Aldovandi, which was made possible by the largesse of Clement VIII.[4] Galileo's condemnation by the Church in 1633 was partly the outcome of his attempt to win over the Church to the Copernican system. It was Galileo's hope that his hitherto friend, Maffeo Barberini, who became Pope Urban VIII in 1623, might act as a patron and protector in his quest to promulgate his new cosmology. But Galileo's condemnation was a forceful indication of the Counter-Reformation Church's suspicious attitude to intellectual innovation, including in the realm of science – thus making it an increasingly unlikely patron of science. Within the Protestant churches there were many devout believers who combined their faith with the promotion of science, but science was not the core business of the churches, and their resources to support it were limited.

Along with the universities and the Church, another possible patron was the aristocracy. There were some instances of this: the patron of the Academy of the Lincei (lynx-eyed) (founded 1603), the first scientific academy, was the Marquess of Monticellis, Fedreico Cesi.[5] Henry Percy, ninth Earl of Northumberland, formed a circle which included astrologers, natural philosophers and mathematicians (such as Thomas Harriott). He also corresponded with Francis Bacon.[6] Aristocrats gave their approval to other later academies, such as the English Royal Society

---

[2] Gascoigne, 'A reappraisal'   [3] Burns, *The Scientific Revolution*, p. 83
[4] Findlen, *Possessing nature*, p. 361   [5] Dear, *Revolutionizing the sciences*, p. 111
[6] Gaukroger, *Bacon*, p. 162

(1660) or the French Academy of Sciences (1666), by becoming members. Few, however, had the means or the inclination to provide what the promoters of the scientific movement needed most: secure employment.

The remaining hope of scientific employment outside the universities was the royal courts. As discussed in the introduction, the state in the period of the Renaissance was still very much conceived of as the person of the ruler. Only to a limited extent could rulers call on the resources of a centralised, bureaucratic state to bolster their position. In a very uncertain world, princes and kings used all available resources, including the achievements of the Renaissance, to glorify their position and their own hold on power. This applied particularly to the unstable world of northern Italian city-states. Many of these had metamorphosed from republics to princedoms as internal disunity had undermined their republican constitutions – as happened, for example, in Florence. The uncertain powers of the Holy Roman Emperor meant, too, that in the German-speaking lands some of the princes sought to consolidate and, if possible, expand their power through the use of patronage, including that of science. Science, then, became one more form of cultural achievement, along with others such as the arts, which could add lustre to the court of a prince or king, thereby consolidating their power. It was a formula well endorsed by that sage observer of Renaissance Italian politics, Machiavelli, who wrote in *The Prince* (1532) that 'a prince ought to show himself a lover of ability, giving employment to able men and honouring those who excel in a particular field'.[7] Note the way in which Machiavelli focuses on the prince as the foundation on which the state rested, rather than later views of the state as an impersonal entity transcending the person of the prince. The result was an emphasis on the personality of the prince that led to the interests of the ruler and state being considered as identical.[8]

The Courts and Astrology and Astronomy

In the quest for positions at court, natural philosophers could build on earlier precedents when rulers had acted as patrons of branches of knowledge that had been frowned upon or marginalised within the universities. For princes, one important field which warranted their support was astrology, with its hopes of planning the future and using such information for royal advantage. Such a preoccupation had medieval roots, as in the case of Charles V of France, who established a college of astrology and medicine at the University of Paris in 1371.[9] The Renaissance's interest in

[7] Eamon, 'Court, academy and printing house', p. 33 [8] Shennan, *The origins*, pp. 25, 40
[9] Moran, 'Courts and academies', p. 252

the occult and new forms of knowledge further encouraged royal patronage of astronomy as a form of control over knowledge to the advantage of the prince. The Sforza dynasty of Milan looked to astrology to reinforce their view of events and their likely outcome. Astrology was also seen as a natural ally of medicine[10] – a viewpoint which followed from the widespread belief that the microcosm of a human being mirrored the macrocosm of the universe. Similarly, Maximilian I (Holy Roman Emperor, 1493–1519) sought to control the astrological data produced by his court astrologers in order to shore up his own political power and to shape public opinion. As far as possible, he controlled the dissemination of astrological information by distributing court-approved printed astrological tables and concentrating astrological activity and associated instruments at the court. Court astrologers were also often drawn from those with a background in medicine. One feature of his politics of knowledge was the patronage of key figures, not only through court appointments but also by exercising patronage at the University of Vienna – an instance of the way in which court and university could not be altogether separated in an age in which universities were frequently dependent on royal favour.[11]

The well-established tradition of court patronage of astrology provided a foundation for court appointments in astronomy. Indeed, the two forms of the study of the heavens were often merged without a clear transition from one to the other. While pursuing his astronomical interests under the patronage of his sovereign, King Frederick II of Denmark, Tycho Brahe delivered an 'Oration on Astrology' in 1574 at the University of Copenhagen. Subsequently, in 1599, he moved to the court of the Holy Roman Emperor, Rudolf II, where, after his death in 1601, he was succeeded as court astronomer by Johannes Kepler. From the emperor's standpoint, the chief function of both Tycho and Kepler was to provide astrological predictions. In response to the supernova of 1604, for example, the emperor consulted Kepler on its possible astrological significance.[12] The emperor, as a patron of diverse forms of knowledge, and particularly the occult, also looked to his court astronomers to add lustre to his reign. The mutual benefit to be gained from this form of patronage was made evident in the way in which Kepler named the astronomical tables (which were the fruit of a lifetime's work on his part and that of Tycho) the *Rudolphine Tables* (1627), after the emperor. Their frontispiece depicted the largesse of the patron in the form of an eagle, the

[10] Azzolini, *Duke and the stars*, pp. 9, 135, 167
[11] Hayton, *The crown and the cosmos*, pp. 3–4, 8, 88–9, 116, 119, 171, 4, 197
[12] Dorn, *The geography of science*, p. 139

imperial symbol, dropping coins from its beak while also offering protection in the form of outstretched wings over Kepler's 'temple of astronomy'.[13] Earlier, Kepler had also publicised his role as a client of the emperor by dedication to him of both his *Optical Part of Astronomy* (1604) and his *New Astronomy* (1609), the latter with a title page proclaiming its imperial connections.[14]

Kepler had the good fortune to step into an established position at court. This had not been true of Tycho Brahe when he served Frederick II before moving to the court of Rudolf II. The uncertain role he was assigned under Danish royal patronage underlines the difficulty of neatly slotting early promoters of the scientific movement into a court culture based on hierarchy and tradition. When Frederick II granted Tycho the island of Hveen as a base for his astronomical enquiries, he did so in the manner of conferring a fief on a feudal lord, with the grant being 'quit and free' for the 'promotion of the *studia mathematices*'. As Robert Westman points out, it was a form of appointment which freed Tycho from the authority of the university or any other institution apart from the monarchy.[15] This gave Tycho considerable freedom and on Hveen he built two observatories, Uraniborg and Stjerneborg, during 1576–96. The feudal character of his position was made evident once again by the fact that these observatories were built by the labour services of the island's inhabitants due to Tycho as their lord. The rather precarious nature of his appointment, however, became evident with the death of Frederick II in 1597. For his successor, Christian IV, did not share his predecessor's scientific interests, and removed his patronage. It was an incident which underlined the limits of royal patronage. For, in an age in which the state was conceived in terms of the person and interests of the individual ruler, a court appointment was subject to the whims of the monarch, and, still more, those who succeeded him.

Galileo and the Medicis

As the examples of Tycho and Kepler indicate, the position of court mathematician (a post encompassing astronomy as a form of mathematics) was part of the retinue of courts eager to gain prestige through the promotion of science. It was to such a post that Galileo was appointed in 1610, at the court of Medici, by Cosimo II, who had succeeded to the Medicean title of Grand Duke of Tuscany in 1609. Galileo had laid a trail to the Medici court with care: in 1605, he had sought permission to

[13] Schiebinger, 'Women of natural knowledge', p. 194
[14] Moran, 'Courts and academies', p. 258   [15] Westman, 'Astronomer's role', p. 124

dedicate a pamphlet on the compass to the future Cosimo II and arranged things so he could act as a tutor to the crown prince.[16] In 1610, Galileo reminded the recently installed prince of his claims as a client by dedicating to him his *Starry Messenger*. This contained an account of his astronomical discoveries using the newly invented telescope – another offering sent to the Medicis. Ties with the Medicis were further greatly strengthened by Galileo naming the moons of Jupiter the 'Medicean stars'. Galileo's campaign was successful, and in 1610 he became a member of the Medicean court.

For Galileo, however, the customary title of Court Mathematician was too restrictive. One of the most constraining features of his earlier career as a university professor of mathematics at Pisa and Padua had been the way in which natural philosophy was kept separate from mathematics. For philosophy, as then practised, depended on logical deduction from key texts, a way of conceiving of knowledge which left little role for mathematics. With the Medicean-conferred title of Court Mathematician and Philosopher, on the other hand, Galileo was able to move more freely between the different branches of science which he had been pursuing,[17] and draw them into a synthesis which undermined the traditional Ptolemaic-Aristotelian world view. The courtly title also went some way to redressing the lowlier status of mathematics,[18] which had an all too tangible reflection in the way in which mathematics professors were paid less than those teaching philosophy.

The move to the Medicean court provided, then, a setting in which Galileo could more fully draw out the consequences of his epochal astronomical observations. It was at the court of the Medicis that Galileo composed his fullest assault on the Ptolemaic system, *The dialogues concerning the two chief world systems* (1632) – a work which led to his trial by the Inquisition in the following year. The advantages accrued by Galileo were, then, considerable, but the Medicis also gained advantages from the arrangement. Though they might have been the Grand Dukes of Tuscany in Galileo's time, this was a title that only dated back to 1569, and the dynasty's origins lay in banking and, before that, as wool merchants. To secure the presence of a scientific figure as eminent as Galileo added lustre to the claims of a dynasty whose place among the major houses of Europe was still a little uncertain.

The Medicis' estimation of Galileo's importance was given a highly tangible form in paying him a salary equivalent to a major court official.[19] Perhaps they were afraid of Galileo being poached by another court:

[16] Westfall, 'Science and patronage', p. 14   [17] Dear, *Revolutionizing the sciences*, p. 104
[18] Biagioli, *Galileo courtier*, p. 156   [19] Biagioli, *Galileo courtier*, p. 105

Wladyslaw IV of Poland wrote admiringly to Galileo of the way in which those who enjoyed the gift of quality won the affection of princes[20] – a sentiment which might have been the preliminary to a recruitment attempt. Not only was Galileo granted a handsome income, but, as Mario Biagioli has shown, he was allowed to assume the dignity of a noble complete with country house.[21] It was a social position which, as in the case of Tycho and the king of Denmark, revealed the extent to which a court scientific post had no clear form, and so assumed that of an elevated member of the social hierarchy. The Medicis also gave Galileo considerable support when he was tried in Rome and his Copernican theory condemned in 1633 – though even the prestige of the Medicis could not outweigh the wrath of Urban VIII. Obligingly, the Medicis overlooked the fact that Galileo had been using his Medici connections to widen his circle of possible patrons to include the papacy.[22]

Court Patronage, Medicine and Astrology

Galileo's support by the Medicis grew out of a tradition of court patronage for astronomy, which, in many cases, had its earlier manifestation in court astrology. As we have seen, astrology had often been linked with medicine, another area of court-supported science. On occasion, pioneering work in medicine was rewarded with a court position, as happened to Vesalius, who dedicated his *About the structure of the human body* (1543), one of the fundamental works of the Scientific Revolution, to the Holy Roman Emperor. This was duly acknowledged with a position as a court physician.[23] Such an appointment often offered an opportunity for scientific enquiry that might go beyond medicine itself. In the late seventeenth century, for example, the Medicis gathered around them a circle of innovative natural historians with the formal title of court physicians – a circle which included Francesco Redi, Marcello Malphighi and Nicolaus Steno.[24] The court physician to Elizabeth I of England, William Gilbert, conducted his original work in natural philosophy, focussing particularly on the nature of magnetism and electricity. His strictures on the scholasticism of the Schools were possible in a court setting, when they would have been frowned upon in the universities. The practice of medicine itself meant the presence at court of members of a knowledge elite who sought to promote the health and wellbeing of those directing the state. This would provide a foundation for proposals

[20] Westfall, 'Patronage', pp. 385–99, 389   [21] Biagioli, *Galileo courtier*, p. 161
[22] Biagioli, *Galileo courtier*, p. 249   [23] Dear, *Revolutionizing the sciences*, p. 110
[24] Boschiero, *Experiment and natural philosophy*, p. 103

for protecting larger sections of the population from disease,[25] as in the case of quarantine.

Like astrology, unconventional forms of medicine, such as the system of Paracelsus (1493–1541), also found a more receptive environment in the courts than in the universities.[26] For Paracelsus's attack on the Galenic four humours system of physiology and his espousal of chemical cures challenged established orthodoxies. The occult nature of Paracelsus's system, with its emphasis on astrology and alchemy, further added to its attractions in courts such as that of Moritz of Hessen (1572–1632), already preoccupied with alchemy.[27] Tycho Brahe's early patron, Frederick II of Denmark, supported Brahe and other natural philosophers, in part because of their Paracelsian interests. Though his successor, Christian IV, withdrew his favour from Brahe, he continued to support other Paracelsians, and provided them with a laboratory.[28]

Part of the attraction of Paracelsianism was that it meshed well with alchemy, another field which tended to be viewed with disfavour in the universities. At a number of courts, however, it was often cultivated in the hope of gaining riches through transmutation of metals, or by developing new and more lucrative methods of manufacture. Along with his expertise as an astronomer/astrologer, Tycho owed his patronage, both by Frederick II of Denmark and the Emperor Rudolf II, in large measure to his interest in alchemy. Alchemy was also given court support by Moritz of Hessen (1572–1632) and other northern princes, reflecting a widespread fascination with the occult.[29] Among such courts with an interest in the occult was that of Elizabeth I of England, who supported the work of John Dee both as alchemist and astrologer.

Mechanical Inventions and Mathematics

Courts were also a more congenial environment than universities for promoting practical skills in areas which today would be described as engineering, particularly as they applied to the needs of war. Interest in promoting natural philosophy and the application of new mechanical techniques often merged. Galileo's early training in mathematics, for example, came from a Florentine court instructor, Ostilio Ricci, who acted as a teacher in a wide range of fields involving the application of mathematics including military fortification, mechanics and architecture.[30] After the telescope became available, Galileo's immediate

[25] Kümmel, 'De Morbis Aulicis', pp. 40–1   [26] Burns, The Scientific Revolution, p. 97
[27] Moran, The alchemical world, pp. 9, 114
[28] Shakelford, 'Paracelsianism and patronage', p. 85
[29] Moran, The alchemical world, p. 171   [30] Gaukroger, The emergence, p. 208

reaction was to seek out practical applications, and especially those from which he might profit. Thus, in 1609, he drew the Doge of Venice's attention to the military possibilities of the new device with which one could 'discover enemy sails and fleets at a greater distance than is customary . . . Also on land one can look into the squares, buildings and trenches of the enemy from some distant vantage point.'[31] Leonardo da Vinci was employed for a time as a military engineer by the Sforzas of Milan, as well as the republic of Florence.[32] Preoccupation with the scientific aspects of ballistics could lead to furthering the study of mathematics. Such was the case at the court of Urbino, which became a centre of mathematics within Italy. Appropriately, it was to the duke of Urbino that Niccolo Tartaglia dedicated his famous works on ballistics, the *Nova scientia* (1537), the 'better to instruct the artillerymen of your most illustrious government in the theory of their art'.[33]

The stimulus provided by the court of Urbino to the study of mathematics extended to a revival of classic Greek mathematics, and particularly that of the Archimedean tradition.[34] This was in keeping with the Renaissance revival of classical texts, which extended to an interest in the scientific texts of antiquity. As the ideological expression of the Renaissance, humanism, the study of ancient texts which largely dealt with the problems of human society, was more suited to the needs of lay intellectuals than the traditional clerical elite. It was, then, slow to gain a foothold in the universities, though this was less true in the more lay-controlled Italian universities than the more clerical northern ones. Courts, then, could act as patrons of the early stages of humanism, and, as it gained increasing prominence at court, it encouraged other forms of learning, including science.[35] The influence of humanism, which may have been reinforced by his time at the court of the Medicis, is evident in Galileo's *The Dialogues concerning the two chief world systems* in the way in which it draws on the classical form of a dialogue of the kind used by Plato. With its strong emphasis on civic virtue, humanism also provided an incentive to link science with practical improvement. In France, the University of Paris strongly resisted the rising humanist tide, but it gained a haven at court. Hence Francis I, having been impressed with the culture of Renaissance Italy, sought to promote humanism by founding a new institution, the Collège Royale. This was dedicated to the study of such patently humanist pursuits as the cultivation of the classical languages,

---

[31] Grove, *In defence of science*, p. 68   [32] Laird, 'Patronage of mechanics', p. 56
[33] Gray, *Science at war*, p. 3   [34] Eamon, 'Court, academy and printing house', p. 32
[35] Findlen, *Possessing nature*, p. 295

but, significantly, it also included mathematics – a discipline that had been reinvigorated by the Renaissance recovery of ancient texts.

Cabinets of Curiosities

Whatever the branch of learning receiving court patronage, whether astronomy, medicine, mechanics or humanism, the goal was the glorification of the prince or monarch by displaying the intellectual capital which he had accumulated. An even more tangible way of displaying the prince's intellectual eminence and wealth was the accumulation of objects and the development of cabinets of curiosities, which were virtually museums in miniature. Exchange of objects between royal collectors was a means of building connections and of shaping the character of such collections and the scientific speculations based on them. Using the collection to provide gifts was an even more potent demonstration of princely status.[36] The vast collections of curiosities amassed by Emperor Rudolf II attracted the attention of scholars, ambassadors and courtiers from around Europe, and grew further as visitors added further curiosities to gain the good will of the Emperor.[37] This Hapsburg preoccupation with turning scientific knowledge to political advantage had been well established before Rudolf II, with his predecessor, Maximilian I, building up a collection of artefacts to add to the lustre of his rule.[38]

Cabinets were highly diverse in character and could include works of art and antiquities together with mechanical marvels. Commonly, however, they included objects of nature, including specimens brought to Europe by the increasing number of voyages beyond the seas. Consequently, such collections acted as a stimulus for the study of natural history.[39] Interest in the works of nature could extend to a royal garden, with some princes displaying their munificence by allowing the garden to be used by the local university. Such, for example, was the case at Ferrara, an early centre for the study of natural history, thanks to the duke allowing access by the university professors to his garden. Cosimo I, Grand Duke of Tuscany, saw the establishment of gardens as an appropriate form of princely patronage, and in 1545 established one both at the University of Pisa and within Florence itself.[40]

---

[36] Findlen, 'Economy', pp. 7, 15, 24   [37] Findlen, 'Cabinets', p. 211
[38] Hayton, *The crown and the cosmos*, p. 200   [39] Moran, 'Courts and academies', p. 263
[40] Findlen, 'Natural history', p. 280

The Spanish Crown and Imperial Science

Natural history, with its specimens drawn from as many different parts of the globe as possible, was one of the most apparent ways in which Europe's increasing contact with the larger world was most evident. This was particularly true at the court of Spain, since that nation had absorbed so much of the New World into its imperial domains. Philip II displayed much interest in the natural productions of his colonised lands, an interest which extended to a natural history of their human population. To further such goals, in 1570 he commissioned a royal physician, Francisco Hernándes (c. 1517–87), to collect and record specimens from Mexico, an undertaking which lasted until 1577. Immersing himself in the local culture, Hernándes learned the local Nahuatl language and drew on indigenous knowledge of healing plants. The outcome was a massive sixteen-volume manuscript work, *On the history of plants of New Spain*, the details of which served as a form of Spanish claim to its newly-won territory.[41] However, such claims were limited by the Spanish obsession with secrecy to pre-empt giving advantages to other rival nations. Consequently, Hernándes's manuscripts had only a very restricted currency until a drastically abridged version was published in Rome by the Academia dei Lincei in 1651.[42]

Hernándes's epic work was followed by an even more extensive attempt by the Spanish monarchy to catalogue the products and possibilities of the New World, the better to control them and exploit them as a source of greater Spanish wealth. This took the form of a series of questionnaires known as *Relaciones Géographicas* sent in 1577 to all royal governors requiring such details as mineral resources, tides, possible harbours and the names of plants.[43] As in the case of Hernándes, particular attention was paid to medicinal plants, a continuing feature of imperial scientific contact. Following a common pattern, such experiments in the colonies were linked with innovations in the metropolis. For in 1575 and 1578 Philip II commissioned similar *Relaciones* within peninsular Spain with a view to collecting statistical information on his kingdom's population and localities.[44]

Empires involve drawing boundaries, and both the Emperor Charles V and his son, Philip II, used the technical skills of cosmographers (those who provided a general description of the Earth) to help clarify the boundaries of their colonial possessions. Empires also require mapping, and to that end Philip II sent out Jaime Juan to ascertain the latitude of

---

[41] Barrera-Osorio, *Experiencing nature*, pp. 17, 134
[42] Cañizares-Esguerra, *Nature, empire and nature*, p. 99
[43] Vogel, 'European expansion', p. 826   [44] Goodman, 'Philip II's patronage', p. 57

particular sites in New Spain and the Philippines and, as far as the available technology allowed, their longitudes.[45] The example of Spain in linking scientific expertise with imperial expansion prompted the rivalry of other nations. Hence that master of the occult, John Dee, used his connection with the court of Elizabeth I to propose applying mathematics to the goals of imperial expansion following the Spanish model.[46] In the event, however, it was not to be the British crown, but rather chartered companies that laid the early beginnings of the British Empire. The British crown, like other states, was increasingly concerned with establishing clear boundaries for their jurisdiction, and hence encouraged scientific and, in particular, mathematical activities relevant to surveying and mapmaking. Court patronage was also commonly extended to those making devices to assist the drawing of such boundaries throughout much of Europe.[47]

In his admiration for Iberian imperial expansion, John Dee, who studied under the Portuguese navigator Pedro Nuñez, would particularly have in mind the achievements of the Spanish crown and, to a lesser extent, that of Portugal in promoting more scientifically based forms of navigation. Without the astronomical skills to calculate latitude, for example, the long-distance navigation of the Spanish and the Portuguese would not have been possible. Such imperatives prompted Philip II to form an Academia de Matematicas, which was so closely connected with the court that it was linked with the royal palace.[48] Information on the best routes for navigation was stored and tabulated in the Spanish Casa de la Contratación (House of Trade), established at Seville in 1503, or the Portuguese equivalent, the Casa da Mina (House of the Indies), founded 1482. The outcome was maps and other information that enabled the imperial fleets to travel more safely around the Iberian empires and the globe more generally. Such a pooling of knowledge, it has been suggested, may have provided one of the models on which the later scientific academies were to be based.[49]

Under Philip II, then, court patronage of mathematics and astronomy became increasingly linked with navigation rather than, as was the case at most European courts, with astrology. As other countries followed the Spanish example, the needs of navigation were to become one of the major incentives for state promotion of astronomy. Yet, Philip II was sufficiently of his time to continue to look to court astronomers for

[45] Goodman, *Power and penury*, pp. 53, 67   [46] Burns, *The Scientific Revolution*, p. 54
[47] Moran, 'Courts and academies', p. 260; Burns, *The Scientific Revolution*, p. 54; Moran, 'Patronage and institutions', p. 170
[48] Pimentel, 'The Iberian vision', p. 21
[49] Barrera-Osorio, 'Knowledge and empiricism', p. 220

astrological advice, though possibly with greater scepticism than other rulers. Interest in the occult also extended to maintaining an alchemical laboratory at the royal palace.[50]

In his plans for welding together his diverse kingdom, science played a significant role. This helps to explain Philip II's interest in the practical applications of science despite some wariness about any possible threats to Catholic orthodoxy from scientific innovations. Large-scale engineering projects were commissioned by the crown, and its centralising role was strengthened by the needs of war and the empire. The crown also provided much of the expertise to conduct such projects. One of the functions of the Academia de Matematicas, for example, was to train engineers rather than importing foreigners.[51] A particular area of interest for Philip II was medicine, with the result that quite a number of physicians benefited from his patronage.[52]

The needs of empire had led to the creation in Spain of a large-scale bureaucracy and, with it, a state increasingly reliant on information.[53] To this extent, Spain was moving towards the form of an impersonal state that transcended the personality of the ruler. There were still traces, however, of a personalised rule. Philip II himself still dealt with a vast amount of administrative detail, prompting one contemporary to remark caustically that 'no secretary in the world uses more paper than His Majesty'.[54] The dynastic, personalised nature of his rule was underlined by the fact that, along with being king of Spain and its empire in the New World, he was also, thanks to his father, ruler of the rebellious Netherlands – albeit a contested one. Philip II, too, had to deal with strong localised traditions, the result of Spain having only recently become a single kingdom. For many of his subjects, then, Philip II was not so much king of Spain but ruler of localised entities such as Castile or Aragon. His patronage of science went some way to promoting a more centralised state, but its influence was too limited to counter strongly the centrifugal forces of the Spanish Empire.

Francis Bacon (1561–1626)

Spain's forms of bureaucracy and administration and its active imperial expansion had an admirer in Francis Bacon, one of the most influential advocates of linking science more closely to the institutions of the state. Ever alert for parallels between the advancement of knowledge and the

---

[50] Cañizares-Esguerra, *Nature, empire, and nation*, p. 44
[51] Goodman, *Power and penury*, p. 261   [52] Goodman, 'Philip II's patronage', pp. 52–3
[53] Soll, *The information master*, pp. 19–20   [54] Elliott, *Imperial Spain*, p. 252

promotion of government, he advocated 'an administration of knowledge in some such order and policy as the king of Spain in regard of his great dominions useth in state; who though he hath particular councils for several countries and affairs, yet hath one council of state or last resort'.[55] The extent of the Spanish Empire was something which Bacon wondered at, writing in his *Of the dignity and advancement of knowledge* (1623) that he had 'marvelled sometimes at Spain, how they clasp and contain so large dominions with so few natural Spaniards'.[56] In his fable *New Atlantis* (1627), depicting an ideal kingdom with its state-maintained research institute, the imprint of Spanish intrusion into the New World is apparent: thus the boat that takes the travellers to the New Atlantis leaves from Peru, and on the island of Bensalem (New Atlantis) the lingua franca between travellers and residents is Spanish.

The Scientific Revolution brought with it hopes of a new ordering of knowledge that would alleviate the lot of humankind. No individual gave clearer and more programmatic voice to such an ideal as Francis Bacon, who was educated for the exercise of politics and served in various roles culminating in that of Lord Chancellor under James I from 1618 to 1621. Though the physician William Harvey dismissively referred to Bacon as writing natural philosophy like a Lord Chancellor,[57] it was Bacon's genius to apply to the promotion of science the lessons he had learned in government. He did so, moreover, with a literary grace in both Latin and English that epitomised the impact of the Renaissance humanism. This was evident, too, in his frequent invocation of the writings and mythology of classical antiquity. In the spirit of Christian humanism, he also saw it as his duty to apply knowledge to the betterment of society rather than simply pursuing science for its own sake.[58]

Drawing on his political experience, the central problem Bacon confronted was the patronage of science. For Bacon was well aware that there was little hope of his plans for the reformation of knowledge the better to serve human needs, unless there were the resources to bring this to pass. A severe critic of the universities and their system of verbal and text-based natural philosophy, Bacon looked to the king to provide the patronage necessary to promote a new form of natural philosophy based on empirical observation. Hence the dedication of his manifesto about the possibilities of new knowledge, *The Advancement of Learning* (1605) to James I and its expanded Latin translation, *De dignitate et augmentia scientiarum* (*Of the dignity and advancement of knowledge*) (1623) to the heir apparent, the future

---

[55] Bacon, 'Valerius Terminus or the interpretation of nature' in Bacon, *Works*, iii, p. 231
[56] Bacon, 'Of the dignity and advancement' in Bacon, *Works*, viii, p. 305
[57] Martin, *Francis Bacon*, p. 172   [58] Cormack, 'Twisting the lion's tail', p. 82

2.1 Portrait of Francis Bacon. Engraved by J. Cochran (1850) from a picture by Van Somer (1626) and published by J. F. Tallis in 1880

Charles I. Without royal patronage it would not be possible to sustain the large-scale organisation and central direction of the sort that he proposed. For Bacon, science was a 'thing of very great size, [not to] be executed without great labour and expense, [and thus] a kind of royal work'.[59] Ever the politician, Bacon drew parallels between a well-ordered state and an efficient system for promoting natural philosophy. Both depended on arriving at laws: in a work advocating the union of England and Scotland he wrote: 'For there is a great affinity and consent between the rules of nature and the true rules of policy; the one being nothing else but an order in the government of the worlds and the other in the government of an estate.'[60]

[59] Bacon, *Works*, iv, 251 cited in Solomon, *Objectivity in the making*, p. 64
[60] Bacon, 'A Brief Discourse touching the Happy Union of the Kingdom of England and Scotland', in Spedding, *Letters and life of Francis Bacon*, ii, p. 90

His orderly mind also favoured greater system and centralisation in both politics and in science. In politics, he favoured not only such measures as the union of England and Scotland, but also the greater systematisation of law drawing on Roman law rather than the haphazard nature of the common law[61] – thus magnifying the scope of the royal prerogative.[62] In science, he advocated a large-scale measure of coordination: firstly, the production of much first hand data (including from observation of the trades) and, secondly, another layer of scientific organisation using such data to derive more general laws about nature, the analogy he employed being the way grapes are turned to wine. Though a cooperative model it was also one which, like the society of which it was a part, hierarchical in character.[63] Such a parallel between statecraft and science was the basis for Bacon's programme of integrating the two so that state power would be increased through its partnership with science.[64] Such a union naturally flowed from his maxim that 'human knowledge and power come to the same thing'.[65]

Bacon depicted in idealised terms the kind of support for science he laboured for in his *New Atlantis*. The image drawn there of a state-supported scientific institution may have owed something to the example of one of the few exemplars of royal patronage of science: the way in which Frederick II of Denmark supported Tycho Brahe, word of which was brought back by James I after his state visit to Denmark in 1590.[66] In *New Atlantis*, an institution known as Salomon's House acted as a research centre with the support of the state. It was hierarchically ordered, with the top officials being arrayed in the manner of the aristocracy, and with a compartmentalised workforce working on different projects but the elite drawing together their results. Salomon's House augmented its researches with information drawn from other societies. Hence 'Merchants of Light' were sent out surreptitiously to make their way to other countries and record any scientific innovations. Though it was state supported, the elite of Salomon's House maintained a degree of independence. They had, for example, an ambivalent attitude to the release to the government of the results of their researches. The Father of the House told the travellers to New Atlantis that 'we have consultations, which of the inventions and experiences which we have discovered shall be published, and which not: and take all an oath of secrecy for the concealing of those which we think fit to keep secret; though some of these we do reveal sometimes to the state, and some not'.[67] It was a provision that prefigured

---

[61] Peter Dear, *Revolutionizing the sciences*, pp. 62–3   [62] Leary, *Francis Bacon*, p. 55
[63] Sargent, 'Bacon', p. 150   [64] Martin, *Francis Bacon*, p. 5; Solomon, *Objectivity*, p. 64
[65] Rees, *The Oxford Francis Bacon. The instauratio magna*. Part II, p. 65
[66] Gaukroger, *Bacon*, p. 73   [67] Bacon, *The advancement*, p. 297

many a dispute between the state and scientists over the ownership of state-supported research.

The island setting of *New Atlantis* reflected reports of growing imperial expansion by Europeans (and particularly by the Spanish) in areas such as the Caribbean and the Americas. An admirer of Roman colonisation,[68] Bacon urged the expansion of Britain's domain, viewing increased military strength as necessary 'for empire and greatness'.[69] Reflecting such goals, he was an investor in the Virginia Company.[70] Bensalem (the New Atlantis) to some extent depended on the materials produced by others. These were brought back either by the Merchants of Light in their undercover operations, or by a regular traffic every twelve years when two ships conveyed manufactures and inventions from around the world. Science, then, was associated in Bacon's mind with the growth of maritime power. This nexus between colonial expansion and the promotion of science was to be a continuing theme in the relations between science and the state.

Bacon, then, mapped some of the ways in which science and the state could both benefit from an alliance – science by material and moral support for its interrogation of nature and its use of its findings for the 'reliefe of Mans estate'[71]; the state by adding knowledge to the bases of power and making science a part of the apparatus of the state. But Bacon had little success in winning over the Stuart monarchs to his cause. The benefits he proposed seemed too remote and too disruptive of tradition to receive royal patronage. For the scholastically trained James I Bacon's proposals were largely incomprehensible: hence his characterisation of Bacon's *Great instauration* as being 'like the peace of God that passeth all understanding'.[72] Bacon himself realised the limitations of the early modern state, with its foundations depending on custom, as a partner for science with its challenge to the intellectual *status quo*. Thus, he wrote in the *New organon* [treatise on logic] (1620): 'political novelty is riskier then intellectual. In affairs of state even change for the better brings fears of disorder, since civil government rests not on demonstration but on authority, consent, reputation, and opinion.' By contrast, in the arts and sciences, 'as in mines, all ought to echo to the sound of new works and further advancement'.[73] Combining political and intellectual change through the state's support of scientific institutions was, then, a hazardous enterprise that required a state with securer foundations and

---

[68] Epstein, *Francis Bacon*, p. 171   [69] Bacon, *Works*, ix, p. 306
[70] Whitney, 'Merchants of light', p. 257 (This piece provides a forceful statement of the case for links between the *New Atlantis* and imperialism.)
[71] Kiernan, *The Oxford Francis Bacon. The advancement*, p. 31
[72] Kiernan, *The Oxford Francis Bacon. The advancement*, p. xxlvi
[73] Rees, *The Oxford Francis Bacon. The instauratio magna*. Part II, p. 147

a willingness to take a more favourable view of change. His hopes of scientific organisation that would parallel the apparatus of government also required a state with a better developed machinery of government and bureaucracy than existed in his time. The effective union of state and science was, then, to require further development of the state as well as science.

Frustrated in the quest for royal support, the older Bacon contemplated the possibility of other patrons. One of the goals of translating his *Advancement of Learning* into Latin (and removing derogatory comments about Catholicism) was the possibility of winning over the Catholic Church, and particularly the papacy, to the role of patron.[74] In 1625, the year of his death, he referred to a range of possible patrons when contemplating how his proposed natural history might be continued. Thus, he referred to it as 'a Work for a *King*, or a *Pope*; or for some *College*, or *Order*'[75] – the last mentioned being a reflection of his hope that a religious order like the Jesuits might devote themselves to his project. No such ecclesiastical patron emerged, and the future of Bacon's vision for science remained dependent on the state, slow though it was to assume the role of patron of science.

The Waning of Court Patronage of the Sciences

In the absence of the impersonal structures of the modern state allowing the systematic institution-building which Bacon favoured, the Renaissance state's support for science largely took the form of individual patronage. By its nature, there was no set pattern for the way in which this patronage worked, its duration or the role to be followed by the client scientist.[76] The individualised nature of the exercise of patronage, however, brought with it considerable flexibility, which was of advantage to the early scientific movement. In a court setting it was possible to be freer to advance novel ideas than in a university with its weight of tradition.[77] Indeed, such innovation was encouraged by court culture with its penchant for the new and its enjoyment of adversarial engagement with orthodoxy.[78] The court setting encouraged the exposition of novel scientific ideas in a manner that made at least some of them accessible to

---

[74] An elaborate presentation copy of this work, *De dignitate et augmentia scientiarum* (*About the dignity and advancement of knowledge*) (1623), intended for Galileo's *bête noir*, Urban VIII, exists in the Pierpont Morgan Library, New York, though there was no evidence that it was actually delivered. Kiernan, *Advancement*, p. lv

[75] Bacon to Fulgentio Micanzio (1625). Anon, *Baconiana Bibliographica*, pp. 198–9. I owe this reference to Prof. Richard Yeo.

[76] Westfall, 'Science', p. 29   [77] Moran, 'Courts and academies', pp. 253–4

[78] Biagioli, 'Scientific Revolution', p. 38

a broad audience. Courts also provided the social legitimation for such scientific ideas to be given a wider credibility. By doing so, the courts encouraged the shift from clerical to lay learning.[79]

But, as science grew in size and complexity, it became less suited to the informal methods of the court, requiring the larger structures offered by the more mature state, such as the academies. The universities, too, began to absorb more of the new sciences, making them more congenial sites for the pursuit of scientific advance. Thus, by the later seventeenth century courtly science was in decline.[80] The case of the Accademia del Cimento (Academy of Experiment), which was nurtured at the Medici court in the 1650s and 1660s, seems to fit the movement from court patronage to the academy, but, on closer inspection, it represented a late manifestation of court support for science. The Accademia was the result of the initiative of Leopoldo, the brother of Grand Duke Ferdinando II, though the latter also gave it his support.[81] The extent of continuing interest in science at the Medician court was apparent in the way in which most of those involved with the Accademia were already members of the court circle. The Galileo debacle, then, had not weakened the Medicis' confidence that the patronage of the sciences rebounded to their glory among the courts of Europe.[82]

Leopoldo's achievement was in promoting collaborative experiments and providing a scientific agenda. He did not, however, provide a constitution nor organise the Accademia as an ongoing institution. Rather, experiments were recorded and eventually published in the *Saggi* (1667). Significantly, they were published without authorial attribution, which both emphasised the corporate nature of the production and the way in which all findings were ultimately to the credit of Leopoldo as patron. His social position helped ensure that the discoveries were authentic, for if they were not, his own standing would be diminished.[83] Keeping his status above the fray of scientific debate may account for the emphasis on experiment and the recording of fact rather than in engaging in scientific speculation.[84] The dependence on Leopold was made evident in the way in which the Accademia's activities ceased when he became a cardinal and departed for Rome in 1667.

The Accademia was, then, a fleeting institution reflecting one of the major difficulties of court patronage of science: its overdependence on the good graces of a single patron. In some ways it was a transitional entity with many of the characteristics of traditional court patronage, but also

[79] Gaukroger, *The emergence* p. 209   [80] Burns, *The Scientific Revolution*, p. 97
[81] Findlen, 'Possessing nature', p. 238
[82] Boschiero, *Experiment and natural philosophy*, p. 20
[83] Biagioli, 'Scientific Revolution', p. 28   [84] Boschiero, 'Natural philosophizing', p. 407

with some elements of the aims of the academies, such as corporate experimentation. It also instanced the phenomenon of courts helping to lend their authority to the foundation of academies, though the longer-lived academies were given greater constitutional form than was the case with the Accademia. Both courts and academies had in common, after all, the fact that they were lay institutions in an age when universities were generally still strongly clerical in character. Court patronage of science had provided a valuable space for the early scientific movement to experiment and speculate. In return, science helped promote the status of princes and kings in an age when learning was being accorded greater prestige – in large measure thanks to the movement of humanism. But the methods of court patronage were too ad hoc to be an ongoing support for the growing scientific movement, which, as the seventeenth century advanced, had to adjust to the needs of a more centralised and impersonal state.

3   Absolutism

Absolutism and Centralisation of Knowledge

The fragile Renaissance state was ill-suited to weather the upheavals of the seventeenth century – upheavals triggered by the increasing intensity of the conflicts between monarchs and overmighty subjects coupled with religious upheaval and strife. In much of Europe, the alternatives seemed centralisation and loss of liberty or continued anarchy. Such was the climate of opinion which shaped the absolute state – a state which attempted to concentrate as much power as possible in the hands of a monarch. One major imperative for such centralisation was the needs of war, which stimulated the development of the state apparatus and, particularly, its tax-raising machinery and the knowledge systems it required.

Yet, though the monarch was the hub of the system, he (and in a few cases, she) ruled with the assistance of an increasingly complex bureaucratic apparatus that gave the state an impersonal character. Frederick the Great summed up this divide between the person of the monarch and the institution of the state when describing himself as 'first servant of the state'. In some ways absolutism was a modernising force, with its break with some of the traditions associated with localised aristocratic rule. Indeed, absolutism could form a natural alliance with some aspects of the Enlightenment, with both sharing a determination to impose order on the unruly practices that had been sanctioned by tradition, often with a patina of religious veneration. But, in an age when communications were poor, there were limitations on the power of monarchs, which meant that absolutism was often the theoretical ideal rather than the actual practice of achieving compromises with local and even feudal interests.

Along with the Enlightenment, the absolute state shared characteristics in common with the developing scientific movement. The movement towards absolutism and the greater veneration for science, in an age when religious certainty had been undermined by the disputes following the Reformation, both reflected the quest for stability after the upheavals of the mid-seventeenth century.[1] The key words of the mathematician and philosopher, René Descartes, in his quest for scientific certainty had been 'order' and 'method', and these underlying imperatives also shaped the form of the absolute state and its attendant bureaucracies. Science was valued by the absolute state for a range of reasons: firstly, it was linked with modernisation and the prestige that accrued to a ruler who promoted such innovation; secondly, it was hoped that science could act as a source of greater wealth in a state the economy of which was framed by mercantilism with its high degree of state control of the economy; thirdly, science or practices akin to it, such as statistical calculation, assisted in the running of the state's bureaucratic processes.

It was France that developed many of the key features of absolutism which were to influence other European states. Under Louis XIV's direct rule from 1661, when he declared that he would do without a chief minister, until his death in 1715, the power of the aristocracy, a rich source of dissent and civil war, was largely tamed. Along with this, the fabric of an effective state bureaucracy was laid down with the palace of Versailles the powerful symbol of the king's central role. In his *Memoirs*, Louis bemoaned the state of confusion that existed when he gained power – a situation he sought to remedy by instituting 'order and clarity in all things'.[2] Reflecting the ever-present demands of warfare, one area where uniformity was imposed in the most literal sense was the insistence that the royal army wear uniforms, along with standardised weapons and pay scales.[3] Achieving greater order and system required knowledge about the kingdom on a scale hitherto unknown and the ordering of such knowledge in ways which paralleled or intersected with science. Accordingly, Louis chose ministers who were capable of assembling vast quantities of data that could serve as a secure foundation for decision making. Of such ministers, the one who took the quest for information furthest was Jean-Baptiste Colbert, who served as the Minister of Finances from 1665 to 1683, this office serving as a natural base for collecting as much data as possible on the resources of the state and the possible revenue to be gained.[4] Among the most important of Colbert's informants were the intendants, an office revived by Louis to ensure royal oversight of the provinces. Reflecting Louis and Colbert's love of system, these officials were required to submit regular statistical reports.

Such growing confidence in the power of figures and written records reflects the increasing pace of commerce and its reliance on financial

[1] Rabb, *The struggle*   [2] King, *Science and rationalism*, pp. 159, 168
[3] King, *Science and rationalism*, p. 244   [4] Soll, *The information master*, p. 7

documentation, a habit of mind which may have helped to shape the conduct of government. Even before Louis's reign, the Duke de Sully (1560–1641), minister of Henry IV, had been advised by the Dutch mathematician Simon Stevin that he should maintain the state's financial records in the same manner as a merchant's accounts.[5] The most basic of all statistics for the purposes of government was the level of population. Colbert laid the early foundations for calculating this by publishing births and deaths for the city of Paris as well as conducting censuses for the colony of Quebec from 1666. Other statistical material collected by Colbert reflected the commercial and military capabilities of the kingdom, such as a register of seamen and the extent of forest holdings, which determined the ability to build ships. Colbert shared with Bacon a love of system and, with it, an impulse for legal reform. Indeed, it was Colbert who was responsible for drawing up during the period 1667–81 France's first centralised legal code. The informing principle was, wrote Colbert, the king's desire to reduce 'into a single body of ordinances all that which is necessary to establish a fixed and certain jurisprudence'.[6] But Colbert's love of system had its opponents, who knew instinctively that knowledge was power and feared the consequences of a break with tradition. Thus, Colbert's grand plan for a national survey of landholding in 1679 was defeated by the privileged orders of the aristocracy and the clergy.[7] Perhaps even Louis XIV had reservations about the power that Colbert's great database gave him – in any case, when Colbert died in 1683 he was not replaced with another minister with comparable powers.[8]

The Formation of the Academy of Sciences of Paris

Given Colbert's determination to base his administration on accurate knowledge, it was a natural extension of his activities to promote a partnership between the state and science. The main vehicle for doing this was the Academy of Sciences, founded in 1666 by a royal decree prompted by Colbert. Colbert's original intention was to create one grand academy to encompass all branches of the arts and science. Such an attempt reflects the centralising impulse of Louis' reign, as well as the desire to keep under control the intellectual life of the nation with its potential for dissent. The grand academy was not achieved, and the Academy of Sciences came into existence as a separate institution.[9] The Colbertian vision was finally achieved, however, in another period

[5] King, *Science and rationalism*, p. 61
[6] King, *Science and rationalism*, pp. 175–6, 179, 190, 273
[7] Scott, *Seeing like a state*, pp. 48–9   [8] Soll, *The information master*, p. 154
[9] Lux, 'The reorganisation of science', p. 192

of centralisation and major change, the Revolution, with the foundation
of the National Institute of Science and the Arts in 1795. Consistent with
Colbert's determination to turn knowledge to the advantage of the state,
the Academy was founded as an arm of the state, with senior members

3.1 King Louis XIV and Colbert visit the Academy of Sciences.

receiving a royal salary and all members the prestige of royal patronage. The insistence on maintaining firm state control over the activities of scientists was underlined by the way in which Colbert banned academies not affiliated with the state, lest they act as centres for dissent and even conspiracy.[10]

To add to the lustre of the newly founded Academy and the glory that thereby accrued to Louis, some scientists of international acclaim were recruited by offering generous payment. Thus, the Dutch Christiaan Huygens, for example, was recruited to act as head of the Academy with a salary higher than that of the French members.[11] Another major international figure whom Colbert recruited was Giovanni Cassini, who headed the Royal Observatory, which, like the Academy, was also founded in 1666 along with the state library and the refounded Royal Garden. It was Huygens who suggested to Colbert that the Academy might embark on compiling a natural history 'very nearly along the lines of Verulam' [Lord Verulam=Bacon].[12] Such a remark was an indication of the way in which the Academy more closely resembled Bacon's vision of a centralised, systematic form of science than the more ad hoc and individual approach of the Academy's British equivalent, the Royal Society, founded in 1660. The Baconian spirit was also evident in the way in which Colbert facilitated the move from the traditional practice of the state sponsoring an individual to acting as patron of a corporate entity.[13] Consistent with this, the Academy's publications in its early days were ascribed to the institution as a whole and its collective investigations, rather than particular individuals, in the manner of the Accademia del Cimento. This practice, however, lapsed with time.[14]

Colbert valued science for its own sake as a pursuit that brought new knowledge into being and promoted the rational values that he saw as framing his administration.[15] Predictably, however, he expected the Academy to contribute to the practical needs of government. One indication of how directly the Academy was meant to serve the king was the way in which it was assigned the task of developing hydraulic systems to enable the fountains of Versailles to function with due splendour. Consistent with Colbert's mercantilist political theory whereby the wealth of the country (and hence its military power) should be controlled for the benefit of the state, particular attention was paid to the trades. Compiling a survey of these was the chief mission of the Company of Arts and Trades

[10] Soll, *The information master*, p. 100   [11] Hahn, 'Louis XIV', p. 199
[12] Soll, *The information master*, p. 292   [13] Stoup, 'The political theory', p. 255
[14] Hahn, *The anatomy*, pp. 26–8   [15] Hahn, *The anatomy*, p. 8

founded in 1693. The Company hoped to combine an encyclopaedic overview of the state of the trades with initiatives to improve their practice, but, in 1699, it was absorbed by the Academy and the task was passed on to it. This merger suggests the growing scope of the Academy along with its willingness to undertake state-sponsored tasks.[16]

Chief among the early major tasks the Academy was assigned was the provision of accurate maps that could serve as the basis for major reforms, particularly those relating to taxation.[17] A related task the Academy was also assigned was the accurate calculation of the distance from Paris to Amiens, using the method of triangulation developed by the Dutch astronomer and mathematician Willebrord Snell in 1617. This had implications both for French topography and science more generally. The calculations assisted in ascertaining the dimensions of the Earth, figures used by Newton. They also helped form part of the basis of a topographical map of France following further calculations of the meridian of France by Cassini de Thury in 1739–40.[18] Much of this work on calculating measurements was done through the Royal Observatory, which remained under the Academy until 1771.

Maps shared with statistics the use of abstract symbols to summarise data and to make comprehensible the on-the-ground problems with which government had to deal. One of the first tasks was to map accurately the borders of the kingdom, the absence of such maps to this point underlining the extent to which the state and its definition was still in the process of development. Reportedly, when the boundaries were established Louis XIV complained that the Academy cartographers had deprived him of more land than he had gained in his conquests.[19] Maps were particularly valuable to the state for military purposes. An instance of this was the way in which France shared with its ally, the Hapsburg Empire, the expertise of Cassini de Thury in connecting the Paris meridian with Vienna in order to make Austria's military maps more accurate.[20]

Concern with the military was an abiding aspect of the Academy's deliberations with the French army and navy supporting aspects of the Academy's activities relevant to their needs. Hence astronomy, meteorology and geophysics received particular attention. Thanks to the way in which the Academy was constituted senior military officials had places reserved for them, giving them a strong voice in the Academy's

[16] Stoup, 'The political theory', pp. 211–34   [17] Stoup, *A company*, p. 49
[18] Langines, *Conserving*, p. 66; Close, *Early years*, p. 11
[19] Soll, *The information master*, p. 301
[20] Widmalm, 'Accuracy, rhetoric, and technology', p. 182

proceedings.[21] Of course, some aspects of the Academy's work, such as the measurement of tides,[22] could have both civilian and naval applications, an instance of the state acting as a patron of science relevant to the larger Republic of Letters as well as to national goals. The Academy was also expected to serve the state's growing bureaucracy. From 1725, for example, the Bureau of Commerce commissioned the Academy to conduct experiments to adjudicate on claims being made by petitioners.[23] Such a role expanded into a continuing involvement with the bureau and the Academy acting as a patent office.[24]

The Academy faced the problem encountered by all state-sponsored scientific institutions – how much autonomy it was allowed in the face of state demands. The academicians accepted that various tasks to assist the state and a certain amount of royal intrusion in their affairs was the price of being a royal academy. They did develop, however, a strong respect for their own rules and practices and used these to ward off royal pressure when it was considered excessive. Such respect for the letter of the law stood in the way of the various attempts at reform, with the result that, on the eve of the Revolution, the Academy had ossified and become overshadowed by tradition. All too faithfully did it follow the precept of one of the ministers entrusted with oversight of the academies that 'one must introduce nothing new'.[25] The links with the traditional order were inscribed in its constitution, with twelve honorary members drawn from the nobility and the requirement that the president and vice-president be drawn from that shallow pool.[26] This assertion of privilege did something to offset the prevailing principle of the Academy, which was so at variance with the outlook of the old regime, that appointment should be on the basis of merit rather than birth.[27]

Nonetheless, the Academy had established that the state could benefit from science and scientists could benefit from the state. The state-constructed constitution of the Academy did serve as something of a barrier to excessive royal intrusion, and Academy membership brought public attention to the work of scientists. The close royal involvement with the Academy brought that most valuable of old regime commodities, prestige, to both the state and the members of the Academy.[28] Such prestige translated into increasing regard for the worth of science in many of the instrumentalities of the state. Science became incorporated into the fabric of government decision-making to the extent that, as

[21] Pyenson, 'On the military and the exact sciences', p. 147
[22] Soll, *The information master*, p. 304   [23] Hahn, *The anatomy*, p. 69
[24] Hilaire-Pérez, 'Invention', p. 930   [25] Rappaport, 'The liberties', p. 253
[26] Gillispie, *Science and polity*, p. 83   [27] Hahn, 'The age', p. 11
[28] Hahn, 'The age', pp. 6–9

Charles Gillispie wrote, 'What was particular to France two centuries ago is that the interactions [between science and government] became regular and frequent enough to be called systematic rather than episodic.'[29] In the course of the eighteenth century it was possible to calculate with increasing accuracy natural phenomena thanks to the invention of scientific instruments such as the barometer and electrometer. This quest for precision, as Andrea Rusnock suggests, was reflected in governmental demands for more accurate vital statistics. Particularly important were statistics on population levels, so critical for determining such basic state functions as taxation and military service.[30] 'There can be no well-ordered political machine', wrote the French intendant, Antoine Aubert, in 1778, 'nor enlightenment administration in a country where the state of population is unknown.'[31]

The Spread of French State-Sponsored Science

The permeation of a scientific mentality and its penetration into the mechanisms of government owed much to the way in which science was institutionalised in a range of state-run educational institutions. Predictably, these were often linked with the military, as was the case with the Artillery School (1679) or the celebrated Royal Engineering School of Mézières, with its rigorous curriculum in mathematics, which was founded in 1748. The latter, however, had its civilian equivalent (founded a year previously) in the School of Bridges and Roads – though the distinction between military and civilian engineering works was always a blurred one given the importance of bridges and roads for moving around troops and material. In 1783, a School of Mines was created, probably in response to earlier German initiatives in this area.[32] Other specialised institutions of higher scientific education included the National Veterinary School at Alford, established in 1765. Like the Academy, such schools also helped to consolidate the importance of merit as a basis for appointment, something which even began to influence the practice of the armed forces. In doing so it eroded some of the ingrained attitudes of the old regimes about the importance of blood. As so often, the spread of science brought with it some unsettling Enlightenment ideas.[33] Government also provided models for the integration of science into manufacture with the foundation of such bodies as the Commission of Powder and Saltpetre, established as a state monopoly

---

[29] Gillispie, *Science and polity*, p. ix   [30] Rusnock, 'Quantification', p. 18
[31] Rusnock, 'Biopolitics', p. 50   [32] Gillispie, *Science and polity*, p. 449
[33] Langines, *Conserving*, p. 419

under the guidance of the great chemist, Antoine Lavoisier. He remained a director until 1791, three years before he was he was guillotined.[34]

The French state also supported the Royal Garden, which dated back to 1635 but had been refounded in 1666, the same year as the foundation of the Academy. Its goals were to achieve a familiar combination of prestige and utilitarian practice – the former including the display of exotic or spectacular flowers and plants and the latter the improvement of French forests and the silk industry.[35] Over the course of the eighteenth century, it became the pre-eminent institution for the support of the natural sciences largely thanks to the eminence of Count Buffon, its director from 1739 to 1788. It helped direct the quest for new and better plants both within France and abroad. The naturalists on Count La Pérouse's ill-fated expedition to the Pacific from 1785 to 1788 (when it was shipwrecked in the Solomon Islands) were given detailed instructions by André Thouin, Buffon's assistant at the Royal Garden, and Buffon himself suggested some possible naturalists to include on the expedition.[36] Among the aims of the expedition, which blended the scientific and imperial, was the hope of finding new plants of economic importance to France.

Scientific instructions for the La Pérouse expedition were also issued by the Academy reflecting the long-standing tradition of the Academy's linking of exploration with science – another respect in which it had helped to establish attention to science as part of the mentality of the age. Early in its history, the Academy had sponsored the voyage of the astronomer Jean Richer to Cayenne in French Guiana from 1672 to 1673. There he accomplished two major scientific tasks: firstly, measuring the parallax of Mars, calculations which, when combined with the readings from other vantage points, enabled the computation of the distance to the planet and, in turn, an estimate of the size of the solar system; secondly, the estimation of the strength of gravity at Cayenne, which he found was less than at Paris, indicating that it was further from the centre of the Earth – a conclusion drawn upon by Newton and Huygens in postulating that the Earth was an oblate sphere.[37]

This expedition helped prompt the most celebrated of the Academy's eighteenth-century expeditions, that of Pierre Maupertuis to Lapland from 1736 to 1737 along with that of Charles-Marie de La Condamine to Peru from 1735 to 1743. The primary goal of the expeditions was to measure the shape of the Earth as near as possible to the North Pole (Lapland) and the

[34] Stapleton, 'Élèves des Poudres', p. 233
[35] Schiebinger and Swan, *Colonial botany*, p. 5
[36] Dunmore, *The journal of Jean-François la Pérouse*, I, pp. xxix, lxxxix
[37] Taton, 'Jean-Dominique Cassini', p. 103

Equator (Peru), in order to establish whether the earth was indeed an oblate sphere, or whether, as the calculations of Jacques Cassini, the director of the Royal Observatory, seemed to suggest, the Earth was flattened at the equator and elongated at the poles. To the chagrin of Descartes' followers, who supported this latter view, the on-the-ground measurements of the expedition bore out the Newtonian model – though there was the consolation that he had drawn on the work of the French Richer. The Academy gleaned further insights into foreign lands, and particularly from the Far East, from the Jesuits who received training from the Academy and reciprocated by sending back information.[38]

French Colonialism

Such expeditions strengthened the view among those directing the French state that the globe as a whole could be understood in scientific terms. When, towards the end of the seventeenth century, Charles Plumier (1646–1704) undertook his natural history of the Caribbean, it was done so at the direction of the state.[39] Such confidence in the possibilities of scientific scrutiny of the globe strengthened the imperial reach of the French state. Hence it sought to bring other lands under the sway of scientific planning, the better to reshape their economies to suit the needs of French commerce. Such an intersection of imperialism and science was most evident in the use of French colonial gardens to produce crops that would be useful to France. It was the French who established the first chain of botanical colonial gardens shaped by scientific principles. In 1716, the first such garden was established at Guadalupe, and others followed at Martinique, Cayenne and Saint Domingue (where there was fruitful experimentation on acclimatising cochineal). The principles of economic botany were pursued with particular vigour in France's Indian Ocean colonies of Île de France (Mauritius) and Île Bourbon (Réunion). Established in 1735, the garden on Île de France was the site of experimentation by Pierre Poivre to challenge Dutch control of much of the spice trade.[40] Exploration in one part of the world could assist French interests in another: thus the thick or 'noble' sugar canes which were brought back from the Pacific by the expedition of Louis Bougainville (1766–9) were planted in the French colonies of Île de France, Bourbon, Martinique and French Guiana (from whence they spread to Brazil).[41]

---

[38] Stoup, *A company*, p. 53   [39] Burns, *The Scientific Revolution*, p. 137
[40] McClellan, *Colonialism and science*, pp. 148–50
[41] Brockway, *Science and colonial expansion*, p. 48

Science, then, came to form an integral part of what James McClellan and François Regourd term the French old regime 'colonial machine' – a system of overseas rule that was a natural extension of absolutism with its strong emphasis on centralised control and the promotion of mercantilist economic controls. Within such a system science, too, was largely controlled from the centre, with metropolitan institutions acting as the primary conduit for the transmission of scientific information.[42] The French state institutionalised expert knowledge into a range of formal organisations. Science and expert knowledge thus became integral instruments of French colonisation. The range of state institutions by which scientific information from around the globe could be assimilated into the machinery of central government was considerable. Among such institutions were the Academy of Sciences, the Royal Observatory, the Royal Garden, the Alford veterinary school and the Royal Society of Medicine. There were also the naval agencies, such as the Marine Academy at Brest and the Dépôt des cartes, plans et journaux de la marine (Naval Depository of Maps, Plans and Journals) established in Paris in 1720, the first of its kind – the navy being an integral part of the 'colonial machine'.[43] Given the importance of the study of natural history and agriculture in the French Empire, the Royal Garden played a particularly important part in constructing a worldwide web of scientific activities that made it a leading centre of research in fields such as botany.[44]

As time went on, however, some local colonial scientific societies were formed, notably the Cercle des philadelphes (Club of Brotherly Love) on Saint Domingue (Haiti) in 1784, an institution raised to the dignity of being renamed the Société Royale des Sciences et des Arts in 1789. Reflecting local needs, the primary concern of this society was medicine, followed by agriculture, with a good deal of attention to natural history and very little to the mathematical sciences. Its acquiescence in the system of slavery on which the local economy was based indicates how little it wished to challenge the local constituted order.[45] In the light of increasing local scientific activity it has been argued that perhaps the notion of a centralised 'colonial machine' is a little too focussed on the metropolitan perspective, and that colonial scientific activities reflect local concerns as well as those of the metropolis. The efficiency of the 'colonial machine' in subjugating far-flung possessions may have looked greater from a Parisian then a local perspective.[46] Nonetheless, the French absolutist state went

---

[42] McClellan and Regourd, *The colonial machine*, p. 23
[43] McClellan, *Colonialism and science*, pp. 289–90
[44] McClellan and Regourd, *The colonial machine*, pp. 25, 60, 86
[45] McClellan, *Colonialism and science*, pp. 239, 268   [46] Roberts, *'Le centre'*

further than any other European state in welding together its scientific institutions at home and abroad into an integrated whole functioning for the service of the state, with both science and the state benefiting from such reciprocity.

Science and the Spanish and Portuguese States

Among other European states based on the principle of absolutist rule, the one that had most in common with the French colonial system was Spain. As we have seen in Chapter 2, it sought early on in the creation of its seaborne empire to draw science into the pattern of influences which shaped colonial policy. The resemblances between the Spanish and the French absolutist states were strengthened with the accession of the French Bourbon dynasty to the Spanish throne in 1700. For both the French and Spanish monarchs, the cultivation of practical technical skills by state initiatives formed an important part of their mercantilist outlook and, with it, an emphasis on increasing national self-sufficiency. Little time was lost in initiating such measures: after the first of the Bourbon kings, Felipe V, acceded to the throne in 1701, a Committee on Commerce and Currency was established to promote manufactures.[47] It was an indication of how reliant technical innovation was on the state that such measures continued to remain central in the reign of the reform- ing Charles III (1759–88).[48] As in France, the absolutist regime was not a fertile environment for innovation from below.

As part of the attempt to draw the threads of empire more closely under metropolitan control, consistent with the principles of absolutism and mercantilism, the Spanish instituted reforms which resembled some well- established French institutions. Belatedly, Spain established a royal gar- den in 1774, but this was re-established and enlarged by the reforming monarch Charles III in 1781. Consistent with the view that science was a force that linked the empire together, the garden was particularly devoted to plants drawn from the Spanish colonies. Many of these plants were brought back by Spanish expeditions to the colonies, among them the Royal Botanical Expedition to Mexico in 1787. One of the major aims of these expeditions was to establish the boundaries of their territories on scientific principles, thus deterring possible rivals. As in France, then, cartography acted as a major link between science and the state. Such expeditions were numerous, reflecting the Spanish crown's determina- tion to collect as much information as possible to strengthen its hold over its colonies and its resources. Fifty-four expeditions were dispatched to

---

[47] Fox, 'Science and government', p. 116   [48] David Goodman, 'Science', p. 18

America between 1735 and 1805, with another five continuing on from America to the Philippines.[49] In the quest for valuable resources the Spanish crown also sent out mining experts to America, both German and Spanish, to hunt for possible sites for greater production of silver and mercury.[50] Further scientific information was gleaned from the responses to questionnaires sent from Madrid to the colonies from 1770. These included queries sent by such institutions as the Royal Botanical Garden and the Royal Cabinet of Natural History, which was established in 1776.[51]

In an attempt to regain a major position among the European powers, the education of the officers of the armed forces was reformed with the result that the Spanish military academies became major centres of scientific education.[52] The navy, as the key to Spain's seaborne empire, received particular attention with the foundation of the Academy of Midshipmen (1717), the Naval College of Surgery (1748) and the School of Naval Engineers (1772). Other state institutions related to the revitalised navy were the Astronomical Observatory with its relevance to navigation (1753), the Hydrographic Depot (1770) and the Hydrographic Bureau (1797). Other indications of the Spanish state's interest in science included the foundation of the Royal Academy of Medicine (1760) and the Museum of Natural Science (1771). The founding of such new institutions extended, too, to the Spanish Empire and its largest city, Mexico City, with the creation of a Royal Botanical Garden there in 1788 along with a Royal Mining College in 1792.[53]

Such innovations encouraged the hope that science could act as a force for the modernisation of Spain and the promotion of greater prosperity through more effective exploitation of resources both at home and in the empire. Consistent with the Bourbon affinity with Enlightenment values, the monarchical promotion of science was seen as providing a justification for its rule, including its imperial domination. The promotion of reason and order through science, then, acted as an alternative mode of justifying rule than that provided by the Catholic Church.[54] Yet, Spain lacked the national commitment to the promotion of science which had been so marked a feature of the French state. One telling difference was that Spain did not establish an Academy of Science until 1834 and so lacked an umbrella body for the promotion of science, which had been so effective in France. Spanish attempts to impose a form of scientific centralisation

---

[49] David Goodman, 'Science', pp. 9–34   [50] Cañizares-Esguerra, 'Nature', p. 57
[51] Bleichmar, 'A visible and useful empire', p. 297   [52] Goodman, 'Science', p. 17
[53] Higueras, *The northwest coast*, p. 9   [54] Cañizares-Esguerra, *Nature*, p. 57

on its Latin American colonies also were limited by the fact that such colonies had well-established local traditions which had taken root in the centuries that had passed between the arrival of the Spanish and the attempted reforms of the eighteenth century.[55] By the early eighteenth century, for example, there were some twenty universities in Spanish America along with numerous religious houses that acted as centres of scientific activity.[56]

Interestingly, Portugal established an academy well before Spain with the foundation of the Royal Academy of Science at Lisbon in 1779. Earlier, a more specialised academy, the Medico-Chirurgical, Botanical and Pharmaceutical Academy, had been founded in 1772. With its neat categorisation of its specimens it illustrated the advances made in natural history under the impact of systems of classification such as that of Linnaeus. As in the case of Spain, the royal gardens and museum (Ajuida Palace Museum), both founded in 1768, served to display the objects both natural and 'artificial' (man-made) collected on overseas expeditions. Portugal and Spain were also alike in the way that military training institutions became influential as centres of mathematical science.[57] The momentum for such innovations owed much to the way in which the Marquis de Pombal had sought to initiate reforms such as greater secularisation of university education. Such reforms consolidated the power of the monarchy by attacking the two orders, the clergy and the aristocracy, which stood in the way of greater absolute power.

In Spain and, to a degree, in Portugal, then, the linkage between absolutism and science existed, but its effectiveness was limited by the demands of traditional vested interests and the lack of commitment to scientific advance on the part of the machinery of government. In France, by contrast, thanks to the range of scientifically informed educational institutions and the public interest in scientific projects such as expeditions, the educated class from whom the bureaucracy was drawn were more inclined to integrate scientific concerns into the normal workings of government. The result was that in Spain scientific initiatives from the monarchy tended to wither away for lack of ongoing nurturing.

Science in the German-Speaking Lands

What of absolutism in other parts of Europe? The Hapsburg Empire and the German princely states adopted forms of absolutism both to strengthen the power of the ruler and to promote projects which would

---

[55] Lafuente, 'Enlightenment', pp. 155–73   [56] Cañizares-Esguerra, *Nature*, p. 47
[57] Frates da Costa and Leitão, 'Portuguese imperial science', pp. 35–53

increase the wealth of their realms. Again, as in France, the link between absolutism and mercantilism is evident with its belief that national wealth depended on the action of the state. Such a linkage again provided an association between science and the state, since science was seen as providing a path to greater prosperity as well as strengthening the modernising credentials of the ruler. In the German-speaking lands the centralising, bureaucratic structures so evident in France had their own form of justification and scientific analysis thanks to the discipline of cameralism. This was a form of political economy dedicated to understanding the workings of the state on rational principles. The goal was for the state to balance competing interests in a way which maximised social harmony as well as prosperity. This was an objective considered as being especially appropriate for the many small states that made up the Holy Roman Empire, out of which was to emerge the future German state.[58]

Reflecting the importance of the universities in the German states, cameralism was conceived and disseminated as an academic study which, in turn, produced the bureaucrats who guided state policy.[59] It began in Prussia at the universities of Halle and Frankfurt an der Oder in 1727, and by the last third of the eighteenth century it had permeated most German universities.[60] The quest by the cameralists for greater prosperity provided, then, the basis for an alliance between the state and some forms of science. Cameralism was, for example, a natural stimulus for the study of population statistics given the mercantilist assumption that the number of people in the territory was an index of its wealth. Such a scientific approach to the study of the subjects of a realm was pioneered by Prussia, which created a statistical office under the reign of Wilhelm I (1713–40).[61]

What was particularly valuable to the German states was increased production of existing industries that contributed to government revenue. Hence mining acted as a stimulus for the development of geology and chemistry, while forestry promoted the study of botany.[62]. One of the major sources of revenue was forestry, and hence large swathes of German forests were organised along principles of order and utility, which matched the outlook of cameralists. The chaos of natural growth was replaced where possible with orderly rows that could be tabulated and maintained in a systematic fashion. Mining also loomed very large in the German-speaking lands and provided the principal non-agricultural

[58] Goodman and Russell, *The rise*, p. 369
[59] McClelland, *State, society, and university*, p. 6
[60] Lowood 'The calculating forester', p. 316
[61] Porter, *Health, civilization, and the state*, p. 49
[62] Wakefield, *The disordered police state*, p. 24

income within the Hapsburg Empire in the early eighteenth century. The German state that did most to place mining on a scientific footing was Saxony, which was spurred into action by the need to counter the bankruptcy brought on by the Seven Years' War (1756–63). The upshot was the establishment of the Freiberg Mining Academy in 1765 with considerable funding from the state, which provided support for its students. The most distinguished of these was Abraham Gottlob Werner, who became professor and curator of the mineral collection at the academy as well as acting as inspector of mines in the Saxon mining service. Such activities conformed well with the cameralist doctrines of promoting self-sufficiency by increasing state revenue. The example of the Freiberg Academy prompted the establishment of mining schools elsewhere in Europe, among them Prague and Berlin in 1770, St Petersburg in 1773 and the School of Mines in Paris in 1783.[63]

The advantages to be gained from science and the prestige that accrued to its patron prompted some of the German states to found academies. The example of France and its form of academy loomed large, with a scientific academy being associated with the programme for the promotion of absolutism that the French state had pioneered. The best known of the German academies was that of Berlin. An early version came into being in 1700 in the form of the Societas Regiae Scientiarium (Royal Society of the Sciences) under the presidency of Leibniz. A stimulus for its foundation was the switch between the Julian and Gregorian calendars, which brought with it some scientific and particularly astronomical discussion. In the authoritarian Prussian state, institutions such as the Academy depended very much on the good will of the monarch. Hence the hostility of Frederick William I, 'the sergeant king', who ruled from 1713 to 1740, contributed to this early academy's demise. With the accession of the Francophile Frederick II ('the Great'), however, another academy came into being in 1746, which, tellingly, had a French title, the Académie Royale des Sciences et Belles-lettres (Royal Academy of Sciences and Literature).

With science much less entrenched in the fabric of the state than in France the functioning of the Berlin academy was much more affected by the personal preferences of the monarch than the French academy. Frederick the Great, for example, regularly intervened in the Berlin academy's proceedings and promoted his own scientific beliefs with their strong Newtonian character.[64] Another Frederick-imposed limitation on the Academy was that he discouraged contact with other scientific academies.[65] This ran contrary to the general ethos of the eighteenth-century academies, which still retained some of the features of the Republic of Letters with the belief in the mutual advantage to be gained from an exchange of information. Such an ideal was indeed proclaimed by Marquis d'Argenson of the Berlin Academy in 1746: 'Literary and learned Europe makes up, so to speak, only one single society, united by a common goal, which is the progress of the sciences and letters.'[66] In terms of funding, however, the Berlin academy had a certain independence since, in return for producing accurate calendars and almanacs, it was allowed the monopoly on their sale.[67]

In Frederick the Great's mind there was a clear analogy between a scientific explanation of the workings of the universe and the functioning of government. Both should proceed from self-evident premises and provide a comprehensive explanation of the whole in terms of coherent and consistent principles. As he wrote, 'A well-run government must follow a system as coherent as a philosophical system can be.'[68] In promoting the order and system that Frederick valued both in science and politics, the Academy played a major part within the body politic. It was its mission not only to assist the monarch in practical matters but also to promote an ethos of Enlightenment and support for the modernising policies of the ruler – a role which distanced the Academy from the universities. So closely identified was the Academy with the policies of Frederick the Great that it became virtually a part of his state apparatus and the bureaucracy which supported it.[69]

Another dimension of state involvement with science under Frederick the Great was the conduct of medicine. Building on the work of his predecessors Frederick supported a *collegium medicum* that insisted on appropriate qualifications for those practising medicine, thus eliminating alternative medical practitioners. The need for university qualifications meant increasing Prussian state scrutiny over the medical schools. Court physicians also reinforced the need for appropriately qualified physicians and for the dissemination of medical instruction based on recent scientific advances. A state-sanctioned medical profession was part of a more general intrusion of the state into the lives of Frederick's subjects.[70]

The nearest rival to Prussia among the German states was Bavaria, where, in the person of the American loyalist, Sir Benjamin Thompson, close links were drawn between science and statecraft – so much so that he was given the title Count Rumford by the Elector of Bavaria, Karl Theodor. In the twelve years he spent in Bavaria from 1784 to 1796,

[66] McClellan, *Science reorganized*, p. 8   [67] McClellan, *Science reorganized*, p. 70
[68] Terrall, 'The culture of science', p. 335
[69] Terrall, 'The culture of science', pp. 338, 343, 347
[70] Geyer-Kordesch, 'Court physicians and state regulation'

Count Rumford endeavoured to apply a scientific approach to a range of policy issues. He also found time for original scientific enquiry using his observations of cannons being bored as the basis for his hypothesis that heat was a form of motion. This was a significant, if originally contested, theoretical position which later helped to lead to the laws of the conservation of energy. In Rumford's mind such theorising was of a piece with the social policy he developed to meet the needs of the Bavarian elector. For his model of society was, like his model of nature, based on mechanical laws that were amenable to experiment. Moving on from his original assignment of reforming the military, Rumford argued for the application of similar principles of uniformity and order in relation to one of the most pressing of social problems: poor relief. Thus, he argued for a centralised system which was compatible with a bureaucratised absolute state, the Munich poor institute. Characteristically, one of the basic foods provided was the scientifically devised Rumford soup. Like so many would-be reformers, he was to discover that remoulding human society along scientific lines was more difficult than advancing science in the abstract, and his poor institute was not a success. Nonetheless, his service was much valued by the Bavarian elector, who found Rumford's scientific approach to a range of problems fitted well with the form of centralising bureaucratic rule to which Bavaria aspired.[71]

St Petersburg Academy of Sciences

With the French example ever before them the German states, then, regarded some state patronage of science as a natural accompaniment to the growth of absolutist rule. Science and centralisation seemed to be natural allies with their common adherence to rational organisation. Science also brought both prestige and the prospect of some economic advantage, though the states were too small to sustain anything like the French scientific establishment with its flagship academy. German scientific culture was, however, to play an important part in establishing a scientific establishment in neighbouring Russia. For the modernising Peter the Great the establishment of what he termed the Saint Petersburg Academy of Science in 1724–5 (which was renamed the Imperial Academy of Sciences and Arts in 1747) was a proclamation of Russia's European status with science a badge of modernity. Symbolically the Academy was funded with the fruits of Peter's great victory in the form of the customs duties and license fees of the Baltic ports which he had captured in war.[72] Where, traditionally, the tsar's legitimacy had

[71] Maerker, 'Political order'   [72] McClellan, *Science reorganized*, p. 78

depended on its alliance with the Russian Orthodox Church, Peter had weakened such a bond by abolishing the Patriarchy and replacing it in 1721 with a department of government, the Holy Synod.[73] Hence Peter's role as reforming tsar came to depend less on tradition and religious sanction, and more on demonstrated success in peace and war with Sweden in raising Russia to great power status through more effective and centralising government – centralisation which came to incorporate patronage of the sciences.

Given the need to create a scientific estate in a country almost totally lacking in a scientific tradition, the Academy initially incorporated other institutions for instilling scientific education, namely a school and a university. Provision was also made for a service to translate key scientific texts into Russian, and its modernising character was also apparent in the way in which it published the first Russian newspaper.[74] There was some debate as to whether Peter would establish a university or an academy but Peter stood firm on an academy.[75] An academy, after all, could be more easily controlled than a university, and the acquisition of one by Russia meant that it was emulating the major absolutist states of Europe. When Peter looked for a model for his academy the one which most appealed to him was the early Berlin Academy, the Societas Regi Scientiarum (Royal Academy of Sciences) devised by Leibniz and founded in 1700 with him as first president. In his extensive correspondence with Peter, Leibniz emphasised the educational function of the Academy, which formed part of the original design in a way that distinguished it from the Paris Academy.[76] This, however, become less pronounced after his death in 1725. When the Academy finally received its charter in 1747, it was split from the university, the latter slowly dwindling away as students were sent to foreign universities.[77]

Leibniz's espousal of a cosmopolitan academy culture also left its mark.[78] Within a year of its foundation the Academy sought its place as a citizen of the Republic of Letters by addressing introductory letters to the Royal Society, the Paris Academy, the Berlin Academy and proposing cooperation with the Royal Society in conducting experiments.[79] The cosmopolitan nature of the Academy was also reinforced by the large number of foreigners among its early membership. German scientists loomed particularly large, along with other nationalities such as its greatest luminary, the (German-speaking) Swiss mathematician Leonhard Euler.

---

[73] Shennan, *The origins*, p. 65
[74] Goodman and Russell, *The rise*, p. 345; McClellan, *Science reorganized*, p. 76
[75] Vucinich, *Empire*, p. 9   [76] Vucinich, *Empire*, pp. 7–8
[77] McClellan, *Science reorganized*, p. 80; Gordin, 'The importation', p. 12
[78] Gordin, 'The importation', p. 12   [79] McClellan, *Science reorganized*, pp. 155–6

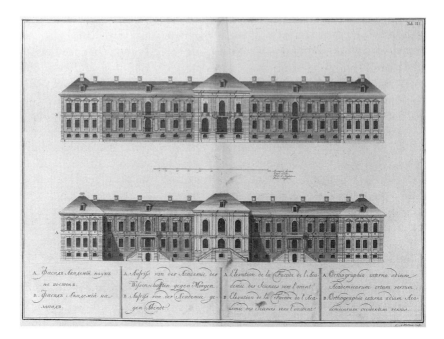

3.2 The Imperial Saint Petersburg Academy of Sciences, 1741

Euler had sufficient standing to be able to bargain with the state, an indication that the Academy had some, albeit limited autonomy. Having left Russia in 1741, Euler was coaxed back to the Academy in 1766 in the early, promising years of Catherine's reign following reassurances that the Academy would be reformed so as to grant greater autonomy for its members.[80]

Along with the prestige gained from publishing new scientific works, the Academy was expected to work on problems relevant to the Russian state. In some cases, this led to further resources being provided to the Academy. In order to compile tables for navigation an observatory was erected – a reflection of Peter the Great's preoccupation with building up Russia's maritime power. As with its counterpart in Berlin, the Academy also used its astronomical observations to compile a national calendar. The Academy also was consulted by the state on a multitude of everyday practical issues ranging from the design and placement of lightning rods and bridge building to the corrosion of ships and the use of coal.[81]

[80] Goodman and Russell, *The rise*, p. 343   [81] Scott-Carver, 'A reconsideration', p. 228

Like the French academy, the St Petersburg Academy oversaw exploration, an important function considering the size and scantily mapped
extent of Russia. One of the major functions of the botanical gardens,
which was also added to the Academy's resources, was to maintain and
catalogue specimens brought back from exploring expeditions. Another
was mapping, a task which was so extensive that in time a geographical
department was established to coordinate the cartographical consolidation of the Russian Empire.[82] Defining the Russian state, then, was
closely linked with exploration of its extent and resources, a task over
which the Academy played a supervisory role.

The first of these major expeditions was the Kamchatka Expedition of
1725–30. One of its major aims was to establish whether or not there was
a strait between Asia and America. It was not able to accomplish this goal,
but it brought back natural history and ethnological specimens, which
formed part of the scientific incorporation of Siberia into the Russian
state. To strengthen further such claims, two geodesists, Ivan Evresno
and Feodor Luzhin, were sent in 1719 to the Kamchatka Peninsula to
conduct a survey. They represented another of Peter the Great's initiatives, the founding of the St Petersburg Academy of Geodesy and
Geography.[83] Like the earlier Kamchatka expedition, the Great
Northern Expedition of 1731–42 was under the command of Vitus
Bering, who, this time, had the satisfaction of establishing that there
was a strait (which now bears his name) between Asia and America –
though he died in the quest. The natural history fruits of the expedition
were extensive, with many being tabulated in the *Flora Siberica* (1747–69)
of the naturalist, Johann Gmelin. The ethnological enquiries of Georg
Steller, which included evidence of the common origin of the peoples on
either side of the Bering Strait, also linked the early science of anthropology with Russia's imperial goals.[84] As in all colonial situations, knowledge
of the local people was a useful preliminary to establishing rule over them.
Such goals were reflected in proposals by the chief cartographer,
Gerhard Müller (a German member of the Academy), for settlements in
Siberia.[85]

Public knowledge of a wider world also was made available through the
St Petersburg Museum, which was founded in 1714. This owed its origin
to Leibniz's conception of the Academy as an integral part of a larger
collection of institutions with the goal of bringing European culture to
Russia – hence Leibniz's suggestion that Peter found a museum. There

[82] McClellan, *Science reorganized*, p. 76   [83] Makarova, *Russians*, p. 33
[84] Dmytryshyn, Crownhart-Vaughan and Vaughan, *Russian penetration*, pp. 104–5
[85] Gascoigne, *Encountering*, p. 118

was a natural association between a centralised state and museums, particularly since, as Paula Findlen points out, a museum helped to establish the public face of the nation.[86] It was possible, however, for states which rejected the absolutist model of government, such as Britain, also to establish museums of considerable stature, such as the British Museum founded in 1753. The foundation of this museum owed much to the bequest of Sir Hans Sloane, a characteristic British mixture of the private and public. The very name of the British Museum suggests the extent to which, in the eighteenth century, museums became associated with national identity, with rivalry between nations promoting the spread of such institutions. The move away from royal or aristocratic cabinets of curiosities to national museums in the eighteenth century underscored the growth of a more impersonal state, with its collection reflecting the character of a nation rather than an individual.

In Russia, as in other absolutist states, the needs of state building and the promotion of the scientific movement intermeshed in fruitful ways. The institution of the academy provided a natural bridge from individual patronage of the Renaissance kind to the more impersonal support of an increasingly centralised state. From the vantage point of the ruler academies had the advantage of being readily controlled and of providing national prestige and practical advice on problems with a scientific dimension. The workings of an academy, with its scrutiny of evidence drawn from direct observation or experiment and use of mathematics, also provided an affirming parallel to the bureaucratic forms which the monarch hoped would be adopted. From the point of view of the scientist academies provided invaluable (if limited) sources of employment and access to the upper levels of government. Eighteenth-century academies also generally maintained links between one another so that being an academician helped to make one a part of the larger Republic of Letters. Inevitably, the abiding issue of scientific patronage – the extent of autonomy allowed to scientists – was a source of tension. This could prompt some academicians to involve themselves in the hazardous enterprise of engaging with court politics, as in the case of Johann Daniel Schumacher, secretary to the St Petersburg Academy of Sciences from 1725 to 1759.[87] A few luminaries such as Euler could threaten resignation but, given the paucity of scientific employment, this was an option open to few.

Along with academies, absolute monarchs also established state-sponsored specialist institutions of higher education to provide experts to serve the state using their scientific skills. These encompassed a wide

[86] Findlen, *Possessing nature*, p. 394   [87] Werrett, 'The Schumacher affair'

variety of institutions generally placing particular emphasis on training relevant to the armed forces, though also including more overtly peaceful pursuits such as medicine, mining and veterinary medicine. In France, where science became more deeply entrenched than elsewhere in Europe, consultation of scientific experts became increasingly a routine part of government. Such attention to science helped to consolidate the state's view of itself as being based on reason and utility, an important ideological underpinning in an age in which the traditional foundations of government – tradition and religion – were starting to wear thin. Yet, there was a fundamental contradiction between the absolutist espousal of scientific values and the actual practice of scientists. Very rapidly, scientists developed their own way of establishing status, which was largely through publication in the journals the academies developed from the late seventeenth century. Such an appeal to merit was at variance with the whole fabric of old-regime society with its hierarchies and orders based on birth and tradition. The association of science with rational processes of government could also prove two-edged when so many aspects of the state, including monarchical rule, were the product of tradition. The old regime largely made its peace with science, but scientific values contributed to the corrosion of the traditional order that reached its climax with the French Revolution.

4   Rivals to Absolutism

Absolutist and Nonabsolutist Forms of Government

Absolutism represented one response to the upheavals of the seventeenth century, but not all states went down that path. In countries where buffers between the individual and the state remained strong, whether through a functioning parliament or a federal constitution, the power of the monarch or central ruler was limited. This was notably the case in Britain, where Parliament's power had been consolidated by the revolutions of the seventeenth century, both that of 1642–60 and of 1688. To some extent it was also true in Sweden, with its traditions of parliamentary rule. Countries with strong provinces and weak central rule, such as the Netherlands and Switzerland, also adopted forms of government which limited absolute powers on the part of the central ruler. As absolutism became more consolidated in Europe, and particularly in France, such nonabsolutist states tended to define themselves in opposition to their absolutist rivals: hence, those features of their government which differed from absolutism tended to be given special prominence.

Absolutism brought with it an elaborate bureaucratic structure that, as we have seen, could act as a foundation for bringing together science and the state. In nonabsolutist states, the scope of the state's machinery of government was often smaller or more localised in character. This had the result that the linkages between science and the state were less programmatic than in absolutist states: in a nonabsolutist state, such linkages often depended on particular alliances between individuals or institutions that were often not incorporated into the mechanisms of government. But whether absolutist or not, by the eighteenth century any state aspiring to great-power status had to take some account of science as a way of providing a route to greater power in peace and war.

The most prominent nonabsolutist European state was Britain, and it was this state that took the lead in the ongoing conflict with France. It was a conflict prompted by great-power rivalry for strategic and commercial dominance. Such conflict also had an ideological quality as a clash between absolutism and forms of government that were regarded as allowing more liberty. Anglo-French conflict was to be a feature of European politics from the end of the seventeenth century until the final British victory over Napoleon in 1815. The battle was largely one of resources, reflecting the amount of national income which could be devoted to warfare. Interestingly, Britain was able to raise almost double the amount of tax that France did,[1] since Parliament provided a way of consulting the population and extracting some grudging measure of consent in contrast to absolutist decrees. French absolutism also had to grant large measures of tax exemptions to the privileged orders to command their assent. But, as the old adage has it, one comes to resemble one's enemies, so that as the Anglo-French conflict continued the British state began to increase the size of its centralised government in ways which had parallels with France. Such a growth in the size of the bureaucracy was to mean more of an incorporation of scientific concerns into the inner workings of government.

The Character of the Royal Society

The contrast between the forms of British and French government was paralleled by the contrast between the British Royal Society and the French Academy of Sciences – even the choice of the term 'society' as against the more formal 'academy' strengthened the divide. The Royal Society was a voluntary body with minimal state funding, while the French was both funded by the state and a part of the governmental apparatus. The reliance on voluntary service was characteristic of Britain, with an elite drawn into the service of the state in various ways, including the conduct of local government and parliament. As the name suggests, the Royal Society was anxious to have the blessing of the king to dispel any suggestion that science was in some ways subversive. Apart, however, from providing a mace and a few incidental privileges, such as the free franking of postage,[2] the Society was expected to raise its own finances. Established with Francis Bacon as its guiding deity, the founding members of the Royal Society hoped to serve the nation by initiating his programme for a history of the trades as a way of adding to the prosperity of society. Bacon's proposals, however, were better suited to the French Academy with its centralised organisation and funding than the more informal and ad hoc methods of the Royal Society. Given its

[1] Brewer, 'The eighteenth-century British state', p. 58
[2] McClellan, *Science reorganised*, p. 30

dependence on the dues of its members, too, the Royal Society had to devote much of its attention to subjects of interest to the social elite, which, as in the case of antiquities, had only a slight connection with science.

Nonetheless, the Royal Society took seriously its role as an advisor to government and was anxious to show the utility of its research for the concerns of government. Early in its history, it began a long and fruitful relationship with the navy. The Royal Society was asked by the Commissioners for the Navy on the best method to improve the supply of timber – an enquiry which did prompt collaborative enquiry of the kind envisaged for the Baconian history of trades.[3] Links with the navy were consolidated by the reforming Samuel Pepys, secretary of the Admiralty from 1673–9 and 1684–8, whose scientific interests extended to ensuring naval officers received an appropriate technical education. Pepys also took an active part in the Royal Society, rising to the position of President, which he held from 1684 to 1686 – a cameo example of how the informal associations that characterised many British forms of government could assist in linking science and the state. As a reformer, Pepys, however, had some sympathy with the French state with its consolidation of power and greater capability of bringing about change in the face of vested interests.[4] Such attitudes were particularly current in the reign of James II (1685–8), until the Glorious Revolution of 1688 firmly entrenched parliamentary rule.

It was Pepys, too, who, along with another government official and fellow of the Royal Society, Sir Jonas Moore, was active in prompting Charles II's government to establish two scientific institutions: the Royal Mathematical School (within Christ's Hospital school) in 1673 and the Royal Observatory, Greenwich in 1675.[5] Both of these foundations reflected the increasing awareness of the need for scientific navigation, thus leading, in the case of the Observatory, to the first purely scientific institution founded by the British state. Sir Jonas Moore served as surveyor-general to the Ordnance from 1669 to 1679 – the department of state entrusted with producing artillery. His background was that of a surveyor, prompting the mathematical interests which were to form part of the working environment of the Ordnance. He also developed an interest in the closely related discipline of astronomy, acting as a patron of John Flamsteed, the first director of the Greenwich Observatory. In the face of royal parsimony, Moore provided Flamsteed with some of his

---

[3] Hunter, 'First steps in institutionalization', p. 21; Underwood, 'Ordering knowledge', p. 105

[4] Hunter, *Science and society*, p. 123

[5] Frances Willmoth, *Sir Jonas Moore*, p. 195; Hunter, *Science and society*, p. 132

4.1 Frontispiece of Thomas Sprat's *History of the Royal Society* (1667) with depictions of Charles II in centre and Francis Bacon on right.

equipment at his own expense. As a former mathematics teacher, Moore was also closely involved in the affairs of the Christ's Hospital school, in which the state took little interest. Along with Lord Brouckner, the founding President of the Royal Society and a member of the Navy Board, Moore also helped initiate experiments within the Royal Society on the physics of gunnery – a subject of great relevance to the navy and the armed forces generally.[6] It was another instance of the way in which the needs of war so often linked science with the state.

The Royal Society also assumed more formal responsibilities, it being the natural place to turn if the state wished to deal with a scientific issue. Though the Royal Society did not control patents in the manner of the French Academy, it was entrusted with exhibiting and registering new inventions before they were granted a royal patent.[7] When the Board of Longitude was established in 1714, Isaac Newton, then president of the Royal Society, gave advice; the Society's president also became an *ex officio* member of the Board. The method chosen to promote a solution to the abiding problem of calculating longitude at sea reflected the nature of British society: no ongoing state position was established, as would have been the natural course of action in an absolutist state. Rather, the Board offered prizes to those who developed their own method using their own devices. It was to prove a successful strategy with the invention of John Harrison's chronometer despite the friction between Harrison and the Board over the reward. The Royal Society, too, was entrusted with oversight of the Royal Observatory from 1710, and from 1767 supervised the publication of its annual reports. Another instance of the flexible links that existed between government and the Royal Society was that when, in 1751, Britain moved from the Julian to the Gregorian calendar, it was Lord Macclesfield as president of the Royal Society who was consulted on points of astronomy.

The Royal Society and Scientific Exploration

The Royal Society was a natural partner, too, in voyages of scientific discovery. The idea that science and exploration with a view to the expansion of empire were natural partners had been a part of the Royal Society's self-understanding since its beginnings. The 1663 second charter had declared in the name of the king that: 'We have long and fully resolved with Ourself to extend not only the boundaries of empire, but also the very arts and sciences.'[8] The first such expeditions were Edmond Halley's two voyages to the South Atlantic in the *Paramour* in 1698 and 1699. These had as their object the determination of what laws governed the variation of the compass. Halley was given command, which led to clashes with a crew discontented with being under civilian command. This may partly explain the dearth of similar voyages until, in 1761,

[6] Hall, 'Gunnery, science', pp. 123–4   [7] McClellan, *Science reorganised*, p. 80
[8] Gascoigne, 'Science and the British Empire', p. 49

British ships were sent to observe the transit of Venus at Cape Town and St Helena. This was very much at the behest of the Royal Society, which sought to be an active member of the international network of scientific societies largely led by France. This scientific transnational alliance sought to measure the passage of Venus across the face of the Sun as a way of calculating one of the most basic of all astronomical yardsticks: the distance of the Earth to the Sun. Lord Macclesfield successfully combined promotion of scientific goals with those of national prestige in urging the British state to play its part in this cosmopolitan venture: it was, he affirmed, a venture for 'the Improvement of Astronomy and the Honour of this Nation'.[9]

A major attraction in becoming involved in observing the 1761 transit of Venus was that it offered the opportunity to act as a sort of dress rehearsal for the transit of 1769 (since transits come in pairs eight years apart). By this time, the Seven Years' War (1756–63) had resolved the Anglo-French conflict over America in Britain's favour. Hence, Britain became interested in the exploration of other parts of the globe, and particularly the Pacific, in the search for national advantage and geographical knowledge. Such voyages took on a scientific character in an age when claims to new territories increasingly had to be substantiated in scientific terms. This area of state activity therefore provided a natural link between government and science. The Royal Society again was the main voice in persuading government to sponsor the second, 1769 observation of the transit of Venus.

After the Halley fiasco, the navy was insistent that command of the voyage be under a naval officer rather than a scientist. This was realised with the appointment of James Cook, who was chosen for his astronomical as well as nautical skills. Cook's *Endeavour* was sent to Tahiti to provide a new vantage point for calculations and thus, it was hoped, to ensure greater accuracy when all the results from around the world were gathered together. Tahiti also provided a natural departure point for exploration of the largely unchartered southern Pacific. On board was the astronomer Charles Green, a Royal Society appointee trained at the Royal Observatory, and the naturalist and future President of the Royal Society, Joseph Banks, who had the means to fund his own voyage complete with equipment and assistants. The success of the Royal Society's representations was evident in the bestowal of 4,000 pounds to aid the voyage, with the administration of the funds being accorded to the Royal Society. Interestingly, this money came from the Crown rather

[9] Gascoigne, *Science in the service*, p. 24

than a vote of Parliament, an indication that expenditures of this kind were not considered part of the normal conduct of government.

Along with Cook's voyage of 1768–71, his third great Pacific voyage of 1776–80 was also largely at the initiative of the Royal Society, reflecting the long-standing fascination with the possibility of a Northwest Passage as a route from Europe to Asia, along with such scientific issues as the behaviour of salt water when frozen. The acclaim of Cook's voyages was to help consolidate a continuing link between naval exploration and scientific investigation, with Charles Darwin's voyage on the *Beagle* (1831–6) being one of its most celebrated fruits. In contrast to such French exploring expeditions as that of La Pérouse's, which carried a full complement of scientists, the British voyagers had few scientists, with many of the observations being carried out by naval officers. The difference underlines the larger size of the French scientific establishment as well as British fears that too many civilian scientists on board would undermine naval discipline – as, indeed, did happen on some of the French voyages.

The Royal Society and Reform of the British State

Despite its voluntary nature, then, the Royal Society was closely connected with government, and became more so as the continued growth of the British state brought with it a larger sphere of government activity. The need to maintain a large fighting force to counter the French, along with the growth of empire, meant a challenge to the traditional hostility to centralisation and the growth of a larger bureaucracy. The unaccustomed experience of defeat in the War of American Independence also prompted moves for reform, one of the most significant of which was stronger state control over the East India Company with the establishment of the Board of Control in 1784. This huge chartered company, then, became more of an arm of state rather than, as hitherto, largely an independent entity; the result was that the British state was more directly involved in imperial affairs. It was a change that brought Britain closer to the French practice of making chartered companies virtually a department of the state. Such changes brought with them a need for a larger bureaucracy and officials who were more likely to owe their position to ability than patronage alone. It also meant the British state had greater need of specialised expertise,[10] which was increasingly true in matters scientific.

It is a mark of the pre-eminence of the Royal Society that its president from 1778 to 1820, Sir Joseph Banks, provided much of the scientific

[10] Eastwood, 'Amplifying the province', p. 293

leadership for the nation during the tenure of his office. The son of a prosperous landed gentleman, Banks inherited the resources to devote his life to what he called the 'scientific service of the public'.[11] No paid official, it was part of Banks's self-understanding that he acted as a voluntary agent, in contrast to academicians in countries like France. Contrasting the Royal Society with such Continental academies, Banks wrote 'we are a set of Free Englishmen, elected by each other & supported at our own expence without accepting any pension or other emolument which can in any point of view subject us to receive orders or directions from any department of Government be it ever so high'.[12] Yet, to achieve his goals, Banks had to enlist the support of the state and, in particular, needed the cooperation of the growing layer of increasingly important civil servants. That the task of coordinating such different branches of government fell to a nonelected private gentleman was an indication of the informal methods that the British state continued to use. Contemporaries recognised the role that Banks was playing as the state's science adviser, viewing it as a natural complement to the more formal mechanisms of government. Hence, Lord Auckland in 1797 jocularly described Banks as 'His Majesty's Ministre des affairs philosophiques' (Minister of philosophical [i.e., scientific] affairs)[13] – the use of the French a telling indication of the perceived association between French statecraft and science.

One such task of coordination that fell to Banks between 1783 and 1787 was overseeing the expenditure of a royal grant of 3,000 pounds intended for an official geodetic survey of England – a natural project for a government anxious to map its territory and its resources. Significantly, the impetus for this came from France, since the director of the French Royal Observatory wished to ascertain, with a degree of accuracy appropriate to astronomical calculation, the distance from his observatory to that at Greenwich. Carrying out the project meant firstly working together with the army, which cleared the ground of the base line on Hounslow Heath – a necessary preliminary for the triangulation calculations. Secondly, Banks needed to draw on the services of the master-general of Ordnance, who provided the skilled engineers. The chief of these, William Roy, was a close ally of Banks, and had been instrumental in establishing the Military Survey of Scotland from 1747 to 1754 following the suppression of the Jacobite uprising of 1745–6 – the beginnings of the British Ordnance Survey and its mapping of the country. The success of the triangulation project helped to lead to an expansion of the reach of

[11] Gascoigne, *Science in the service*, p. 22   [12] Gascoigne, *Science in the service*, p. 31
[13] Gascoigne, *Science in the service*, p. 14

the central government with the creation of the Ordnance Survey in 1791, which expanded the surveys of the country and, with them, the production of the Ordnance Survey maps.[14] By such indirect methods as its involvement in the triangulation project was the influence of the Royal Society expanded as the British state found increasing uses for science.

Imperial Botany

Other institutions apart from the Royal Society became more prominent in linking science with state instrumentalities in the last few decades of the eighteenth century, when the role of government was expanding to deal with the increasing demands of war and empire. In 1773, the Royal Gardens at Kew were transformed from a royal pleasure garden to a botanical research institute under the directorship of Sir Joseph Banks. Such a transformation was an indication that Britain was beginning to devote more attention to the economic possibilities of botany both at home and abroad. Over time, Kew was to become the centre of a worldwide network of botanical gardens scattered throughout the empire. Even in the late eighteenth century, Kew's influence was expanding thanks to its links with an increasingly geographically disparate range of colonial gardens: St Vincent in the West Indies (1764), Jamaica (1775), Calcutta (1787) and Penang (1800).[15]

The move to increasing state involvement in the possibilities of imperial botany and potentially lucrative botanical exchanges underlined the way in which the British state was becoming less reliant on chartered companies (and particularly the East India Company) to run imperial affairs. It was, indeed, following the lead of France in starting to construct a state-run 'colonial machine', which included attention to science, especially in its more utilitarian aspects. One aspect of this machine was, from 1784, the Board of Control, which could manoeuvre the Company in directions favoured by the state. Thanks to a combination of both the Board and the Company, the Calcutta Botanic Garden was founded in 1787. Significantly, its founding director, Robert Kyd, lamented 'the shame of being 20 Years behind our neighbours [the French] in everything of this kind'.[16] As Drayton suggests, the revitalisation of the French Empire from 1763 that followed France's defeat in the Seven Years' War was to help spur the reforms that took place in the British Empire from 1783, following defeat in the American War of Independence.[17] British interest

---

[14] Widmalm, 'Accuracy, rhetoric, and technology', p. 179
[15] McClellan, 'Scientific institutions', p. 102   [16] Drayton, *Nature's government*, p. 112
[17] Drayton, *Nature's government*, p. 11

in imperial botany was highlighted by the famous *Bounty* voyage, which set out in 1787 with the mission of transplanting breadfruit from Tahiti to the West Indies. Though foiled, the task was finally accomplished by William Bligh with the voyage of the *Providence*, 1791–4. Both voyages involved close cooperation between the navy and the Royal Society in the person of Joseph Banks, an indication of how scientific concerns were increasingly being incorporated into the state bureaucracy.

The East India Company still played a role in disseminating scientific information as an incidental outcome of its commercial activities. Along with the linguistic research for which it became so famous, the Asiatic Society of Bengal, which was founded in 1784, published reports based on information supplied by Company employees on the full gamut of natural history – flora, fauna and the human population.[18] Imperial control meant mapping of territories, and the beginnings of this were undertaken in India, with much help from Indian employees, in the late eighteenth century. The chief cartographer was James Rennell, appointed by the East India Company as Surveyor-General of Bengal in 1767, in which capacity he produced an atlas of Bengal in 1780 and a map of India in 1783.[19] The latter was to have almost iconographic significance, creating, as Matthew Edney stresses, an image of the subcontinent in the mind of the British imperial classes.[20] Though an employee of the East India Company, Rennell had well-developed links with individuals who were close to government, like Joseph Banks, and in 1791 he was awarded the Royal Society's Copley Medal.

Rennell's work made India seem more accessible to British interests, prompting Joseph Banks to ponder what benefits could be obtained for Britain through the cultivation of botanical transfers. Presciently, he suggested the possibilities of transplanting tea from India to China, thus saving Britain the bullion the Chinese insisted on for the purchase of tea. His memorial on the subject stresses the way that such an instance of imperial botany could work to Britain's advantage as well as strengthen commercial ties between Britain and India: 'A colony like this [India] . . . seems by nature intended for the purposes of supplying her [Britain] with all raw materials and it must be allowed that a colony yielding that kind of tribute, binds itself to the mother country by the strongest and most indissoluble of human ties, that of common interest and mutual advantage.'[21] Tellingly, the memo was written in 1788, a year after the foundation of the Calcutta Botanic Garden, which, in conjunction with the Kew worldwide network, was the site for early experiments in

[18] Prakash, *Another reason*, p. 52   [19] Arnold, *Science, technology and medicine*, p. 4
[20] Gascoigne, 'Joseph Banks, mapping', p. 159   [21] Kumar, *Science and the Raj*, p. 72

transplanting Chinese tea to India. Banks's ruminations were an indication of the way in which the possibilities of empire were becoming more
a matter of concern after the state became more directly involved in ruling
India with the establishment if the Board of Control in 1784. This
brought with it an increasing linkage between science and the state,
particularly in the form of imperial botany.

Ireland and the Beginnings of Political Arithmetic

England may have only gone down the path of a more centralised and
bureaucratised state with great reluctance but, as ruler of Ireland, it had
employed more absolutist methods. This was particularly true after the
crushing of Irish resistance by Oliver Cromwell in 1652, which meant, as
William Petty, the virtual founder of political arithmetic, put it in 1662,
that 'Ireland is a white paper.'[22] Such was the setting for Petty to develop
some of the techniques of measurement based on scientific principles,
which provided the state with the tools for more effective and more
controlling government of its subjects. As Patrick Carroll puts it,
Ireland provided a setting for the emergence of a 'data state' in the '"living
laboratory" of English science and government'.[23] Developments in
Ireland, then, helped to shape subsequent policies in Britain itself.
As England's first colony, too, some practices in Ireland were to be
transplanted into the larger empire as Britain became more involved in
the direct government of its overseas colonies. Symptomatically, Petty's
interests in Ireland later extended to the colony of Pennsylvania.[24]

Ireland provided the setting for the emergence of political arithmetic,
an attempt to frame social problems in mathematical language, or, as
William Davenant put it in 1698, 'the art of reasoning by figures, upon
things relating to government'.[25] Its pioneering exponent and coiner of
the term, William Petty, regarded post-conquest Ireland as a blank canvas
which he could reconstruct using abstract and, to his mind, rational
principles. His pathbreaking Down Survey of Ireland of 1655–6 was
a cadastral survey mapping the land with a view to its apportionment
among the conquering English. It also allowed for greater control by
government by linking the administrative units with the blocks of land,
while also distinguishing potential resources. The scientific manner in
which it was carried out helped to earn him membership of the Royal
Society in 1662, and in 1683 he was to be among the founders of the

---

[22] Carroll, *Science, culture and modern state formation*, p. 81
[23] Carroll, *Science, Culture and modern state formation*, p. 26
[24] Burns, *The scientific revolution*, p. 127
[25] Rusnock, 'Biopolitics: political arithmetic', p. 49

Dublin Society for the Improvement of Natural Knowledge, Mathematics and Mechanics – an indication of the extent to which he regarded his work as shaped by scientific concerns. A disciple of Francis Bacon, he regarded political arithmetic as the application to the problems of government of Baconian principles. Thus, he noted approvingly the way in which Bacon had pointed to 'a judicious Parallel . . . between the Body Natural and Body Political'.[26]

For Petty, what efficient government required was the tools that science was providing. The state should base its policies on 'foundations [which] are sense & the superstructures mathematicall reasoning for want of which Props so many Governments doe reel & stagger, and crush the honest subjects that live under them'. Echoing the practice of scientists using instruments, he declared his intention 'to express my self in Terms of *Number, Weight, or Measure*' rather than 'superlative Words'.[27] Much of his life was devoted to diagnosing the problems of the Irish economy and the reasons for its backwardness. Such considerations prompted further statistical enquiry, leading to a census of Ireland based on mathematical approximation in 1659, and a full mapping of the island in 1673. It was a mark of his tendency to contemplate solutions based on criteria remote from the lived experience of individuals that he considered an exchange between the bulk of the Irish population with English as a way of promoting productivity in Ireland.[28]

Petty's friend, John Graunt, developed Petty's political arithmetic to provide the beginnings of a study of demography in his *Natural and political observations . . . upon the Bills of Mortality* (1662). This was a line of investigation that was probably suggested to Graunt by Petty, who, wrote the biographer, John Aubrey, had 'a hint from his intimate and familiar friend'.[29] Like Petty, Graunt sought to establish patterns in the data on births and deaths as a potential tool for states to use since 'by the knowledge whereof, Trade and Government may be made more certain and Regular'.[30] By drawing conclusions from an array of empirical data, Graunt saw himself as following the Baconian agenda.[31] His work was presented to the Royal Society to which Graunt was elected in 1662, and was built upon by one of the Society's most eminent figures, Edmond Halley.[32] For, in response to a request from the government of William III, Halley devised methods to calculate annuities as an aid to assessing the contracting of loans. The results of these computations were embodied in his *Estimate of the degrees of the mortality of mankind* (1693).[33]

[26] Porter, *Trust in numbers*, p. 19   [27] Hunter, *Science and society*, p. 121
[28] Hunter, *Science and society*, p. 122   [29] Glass, 'John Graunt', p. 65
[30] Hunter, *Science and society*, p. 121   [31] Webster, *The great instauration*, p. 444
[32] Glass, 'John Graunt', p. 77   [33] King, *Science and rationalism*, p. 175

Though Petty's and Halley's forms of political arithmetic were commissioned by government, the characteristic British pattern was for those developing the field to work independently (as Graunt had done).[34] For example, 'medical arithmetic', the application of political arithmetic to the occurrence of disease, was pioneered and named by the physician, William Black. Another impulse for the development of political arithmetic was the growth of life insurance,[35] with the work of Richard Price (1723–91), published in the Royal Society's journal, both assisting the development of that industry and contributing to the infant discipline of demography. As a radical dissenting minister, he was well removed from government circles. Such detachment from the workings of government was an indication of the limits of the purview of government that was part of the self-image of eighteenth-century Britain (often heightened by invidious comparisons with France). Political arithmetic may have been largely developed in England, but its use as a method of social engineering of the sort that Petty had proposed in Ireland was regarded with suspicion. For much of the eighteenth century, England still remained committed to localised government, and equipping the central state with statistical tools prompted caution. Proposals for a census in 1753 were rejected as favouring an expansion in government power; likewise, a measure for the mandatory recording of births, marriages and deaths was rejected for similar reasons.[36] Not until Britain was subject to the dislocations brought about by war and industrialisation was the central state granted greater powers and, with them, more access to the statistical tools to make its form of government more effective. One indication, however, of the direction of change was the way that political arithmetic in the form of tabulated data on the statistics of trade was an integral part of the deliberations carried out by the Committee for Trade and the Plantations [Colonies] of the Privy Council from its establishment in 1784.[37]

Eighteenth-Century Britain and the State: an Overview

From the time of the restoration of the monarchy in 1660 to the eve of the French Revolution of 1789, the British state gradually evolved from one which left, as far as possible, government to the localities, into one which was beginning to develop more centralised institutions. Changes in the character of the state also meant changes in the type of sciences supported. In the early period, mathematical sciences received particular

[34] Rusnock, 'Biopolitics: political arithmetic', pp. 51, 53   [35] Carroll, *Science*, p. 90
[36] Johannisson, 'Society in numbers', p. 349   [37] Gascoigne, *Science in the service*, p. 12

attention since they were of special relevance to the army and navy, which was the primary responsibility of the central state. Their pre-eminence also reflected the prestige they were accorded following the work of Newton. Accurate navigation was vital both to the Royal and the merchant navy, something which was recognised by the state with the creation of the Royal Observatory and the Royal Mathematical School (to train seamen).

Mathematicians were also commonly employed in the Board of Ordnance, which produced canons, and their skills were put to use in the surveying and mapping that ultimately led to the foundation in 1791 of the specialist division devoted to producing maps. Mathematics also underlay the development of political arithmetic, a field in which the British made considerable contributions – though the fruits of it were only sparingly used by the British state in this period, reflecting the fear of giving central government too much power. In the decades preceding the French Revolution, more attention was given to natural history and especially botany, reflecting the growing standing of these fields with the devising of rigorous systems of classification. Such fields became of more concern to the state as the economic advantages to be gained from the pursuit of botany on an imperial scale were demonstrated. Britain may have sought to keep the central government as small as possible, but this did not preclude the support of a range of scientific activities.

Scientists benefited, too, from the prestige which the state helped to confer on their work. The Royal Society may have been a voluntary body, but it was one with the close connections with government natural to an oligarchy in which a small elite dominated the major institutions. Membership of the Society or support by it in such forms as publication or awards, then, carried considerable social cachet (if not payment). Ideologically, too, science was claimed as a support for the purportedly rational form of government the British enjoyed. Even Newton's cosmology was enlisted to support the Hanoverian ascendency and its form of constitutional monarchy. Hence, in 1728 (a year after Newton's death) John Desaguiliers published an allegorical poem with the title system of the world, the best model of government. Science, then, was valued by the British state both for the ideological authority it bestowed and for some of its practical advantages. Where the British state provided only limited support was in the provision of employment, with the result that science was often a pursuit for leisured gentlemen.

Science and the State in the Netherlands, Switzerland and Sweden

Britain was not alone in eighteenth-century Europe in resisting the strong tide towards absolutism. The ascent of the Dutch King William to the British throne after the revolution of 1688 drew Britain into an alliance with the Dutch and other allies in resisting the expanding power of France. Given that the Netherlands then consisted of a collection of equal provinces, its constitution was naturally unconducive to absolutism. It also shared with Britain Protestantism and an economy very dependent on overseas trade and the development of imperial power. Like Britain, for much of the century the Netherlands relied on chartered companies to a large degree for the maintenance and expansion of its empire. The mighty Dutch East India Company, with its lucrative trade in spices, principally obtained from what is now Indonesia, did encourage interest in natural history[38] – particularly if it took the form of economic botany with market potential. The fact that a number of the Company's employees shared such interests is evident in the foundation of a Batavian (Jakartan) Society in 1788. In telling contrast, a Dutch national scientific academy was not founded until 1808, when the Royal Institute of Sciences, Literature and Fine Arts was established during the period of French occupation. The dominance of the East India Company meant that it took on functions that in other countries belonged more squarely to the state. The interests of the Dutch state and the East India Company were, however, so entangled that it is difficult to draw a clear line between them even before the Company was nationalised in 1796.

The late foundation of a scientific academy in the Netherlands does not mean that there was not considerable attention to science. The Netherlands was effectively a federal system, with each province jealous of the greater ascendancy that another might gain by being chosen as the site for a national academy. It also had a strong system of universities, including, notably, the University of Leiden. This boasted a botanical garden going back to 1587 that acted as a place of experiment for botanical specimens drawn from around the Dutch Empire. Similar considerations also help to explain the late foundation of a national academy in Switzerland, another country with a federal constitution and a strong university system.[39] There the Swiss Academy of Sciences was not founded until 1815.

The federal nature of the constitutions of the Netherlands and Switzerland, then, were a brake on any moves towards absolutism, and

---

[38] Burns, *The scientific revolution*, p. 139   [39] McClellan, *Science reorganised*, p. 123

also inhibited the development of scientific centralisation. In Sweden, as in Britain, absolutism was kept in check by strong parliamentary traditions. Parliament held particular sway in the period from the death of the Swedish king, Charles XII, in 1718, following defeat by the Russians, until the *coup d'état* of Gustavus III in 1772. From 1739, the Swedish parliament (Riksdage) was dominated by the party known as the Hats (as opposed to the Caps), which represented aristocratic landholding interests.[40] The year 1739 was also the foundation date of the Royal Swedish Academy, which was linked to the Hats' objective of promoting mercantilist policies to build up Sweden's self-reliance. This had the considerable scientific support of Linnaeus, one of the founders of the Academy, who developed the most influential system of classification for natural history. The need for Sweden's self-reliance was heightened by its loss of colonies by the mid-eighteenth century, prompting a determination to cultivate useful plants at home. Linnaeus's disciples roamed the globe in the search for such plants with the support of the Swedish Academy. The result was the creation of a worldwide network of both Swedish-based and locally-based collectors who promoted Linnaeus' hopes of restoring to Sweden some of the wider influence it had lost through its military defeats.[41]

Though the Academy was modelled on the Royal Society with its independence from government,[42] the involvement of the Swedish Academy in building up such networks linked it closely to the state. In return, it was rewarded with a fair degree of economic independence by being granting the rights to collect the revenue from the publication of almanacs.[43] The Academy also played a major role in another aspect of mercantilist economic policy, the compilation of national statistics. These enabled a form of political arithmetic to measure the economic health of the country, with the level of population being considered particularly vital. With the active support of the Academy, an Office of Tables was created to coordinate returns from parishes of births, deaths and marriages. This was linked with a census in 1749, the first in Europe.[44] The power that such statistics gave the state was, however, contentious, and in 1772 parliament withdrew its support for the Office of Tables.[45] Other state instrumentalities also played a role in promoting scientific investigation: the Board of Mines made Sweden a leading centre of minerology, and the Mint sponsored a laboratory under the direction of

[40] Eriksson, 'Commentary', p. 33
[41] Reidy, Kroll and Conway, *Exploration and science*, p. 49; Sörlin, 'Ordering the world'
[42] Goodman and Russell, *The rise*, p. 247   [43] Sörlin, 'Ordering the world', p. 69
[44] Porter, *Health, civilization, and the state*, p. 53
[45] Johannisson, 'Society in numbers', pp. 357–9

prominent chemists.[46] When, in 1771, the despot Gustavus III seized control and minimised the role of parliament, however, these scientific initiatives lost much of their momentum.[47]

Sweden, then, is another example of a nonabsolutist state which sought to turn science to its advantage. Science could work with absolutist states such as France, Prussia and other German states and Russia, as well as nonabsolutist states such as Britain, the Netherlands, Switzerland and Sweden. In the eighteenth century, as in centuries to follow, there was no one model for the interaction of the state and science. Yet different forms of state could encourage particular aspects of scientific activity. In Sweden, the need for economic self-sufficiency following the loss of colonies and military defeat promoted natural history and, particularly, botany with the support of the great Linnaeus. Political arithmetic was popular for a time in Sweden as an aid to economic advance but waned with growing political disquiet. In Britain, too, political arithmetic made considerable strides, but there was resistance to linking it too directly to the power of the state. Absolutist regimes had far fewer inhibitions in using political arithmetic, having fewer fears about the encroachment of the state on the individual's liberties. Absolutist and nonabsolutist states were both keen to promote sciences that advantaged their armed forces or maritime trade, as instanced by the near simultaneous foundation of royal observatories in France and Britain. It was in France, however, that the marriage of the state with science went furthest in the eighteenth century, creating models that other nations often sought to emulate in ways consistent with the character of their state.

[46] Goodman and Russell, *The rise*, p. 319
[47] Widmalm, 'Instituting science in Sweden', p. 240

5   Revolution, Reaction and Reform, 1776–1850

The revolutions of the late eighteenth century, the American, French and Industrial, both built on and destroyed the existing forms of the state. Revolution weakened traditional sanctions for the state based on tradition and religion, prompting many states to look for new forms of legitimacy based on reason. State support for the cultivation and dissemination of science provided a particularly tangible manifestation of such rationality. The French Revolution, particularly, also fostered the growth of the nation-state, the merger of a particular people and their culture with the political forms of the state – a conjunction which was to fuel other movements for separate nation-states in the centuries to follow. The ideal of the nation-state brought with it an emphasis on the elimination of barriers between the people and the state – hence, for example, the downplaying of the old divisions in France created by the provinces, and their replacement by departments based on utilitarian principles and rational design.

With the belief in the overarching fusion of the state with the nation went an emphasis on uniformity, one important example of which was the establishment of common units of measurement – an area of national life where science played an important role. A nation-state brought with it, too, a greater emphasis on the way in which the state responded to the needs of those who were its citizens, rather than, as of old, their subjects. All this meant further growth of the 'data state', and the collection of information organised along systematic lines to serve both the needs of the state and, in some instances at least, its people. Cutting through the debris of ages meant greater centralisation of the government apparatus that served the state, the effect of both political and industrial revolutions. Britain might have rejected political revolutions, but the Industrial Revolution brought with it the need for centralising such basic aspects of life as the exact time, a product of the railway age, which demanded one standard time rather than a multitude of different locally set times.

The French Revolution

France under the old regime had, as we have seen, absorbed into the body politic a considerable respect and role for science. There was, however, a basic contradiction in the ideological supports of the pre-Revolutionary order: on the one hand, its institutions were increasingly coloured by a scientific outlook, but on the other, its basic foundations rested on tradition, religion and hierarchy. For the revolutionaries, science and its possibilities offered one path forward in the creation of a whole new epoch of human history based around rationality and reason. It was to such ideals that the impassioned reformer, Nicolas de Condorcet, mathematician and last perpetual secretary of the old Academy of Sciences, turned before he himself became a victim of the revolution in 1794. Tellingly, Condorcet couched some of his remarks on the role of science in the form of a response to Bacon, who had become one of the icons of the Enlightenment in France as well as Britain – hence the title of his work, *Fragment on the New Atlantis, or combined efforts of the human species for the advancement of science* (1793).

For Condorcet, the logical result of the sort of prominence which Bacon had accorded science in *New Atlantis* was a new society purged of kings – even though Bacon had not himself envisaged this. 'Bacon wrote at a time', remarked Condorcet, 'when events had not yet determined whether the inevitable fall to which kings had been condemned by reason would be the peaceful work of enlightenment or the rapid effect of the indignation of peoples freed from deception.' Ideally, what Condorcet hoped for was a form of Academy which would be independent of the state since 'it would be too dangerous to allow any authority to introduce itself into an empire where truth must hold undivided sway'. Nonetheless, science deserved the support of the state that would 'establish relations between the public power and a free association'.[1]

If revolutionaries such as Condorcet held such high hopes for an alliance between science and the state, why, then, was the Academy of Sciences abolished in 1793 (along with other royal academies)? The Revolution was suspicious of all institutions, whether churches, universities or guilds, that held special privileges which stood between the citizen and the controlling arm of the state. Such was the perception of the Academy as a body that claimed for itself rights not enjoyed by other citizens, something that ran counter to the egalitarian ethos of the Revolution. Such a critique was heightened, ironically, by the fact that the alliance between the Academy and the monarchy had been too

---

[1] Baker, *Condorcet*, pp. 284, 298

successful, so that the Academy was identified with unpopular royal policy.[2] After all, it was the *Royal* Academy, and the stripping away of devotion to the person of the king rather than the abstract entity known as the state was a primary goal of the Revolution. Elimination of the attribute of 'royal' and replacing it with 'national' (as in National Institute) was one of the ways in which the Revolution further promoted the understanding of the state as an impersonal entity divorced from any one individual. For even in its choice of institutional names the Revolution sought to remind its citizens that their primary loyalty was to the state that oversaw all aspects of the life of the nation. Hence, for example, the Garden of the King was renamed the National Museum of Natural History in 1793.

Such was the recognition accorded to science as a part of French national life, however, that amidst all the hubbub of the Revolution, an alternative to the Academy was established just two years later in 1795. It took the form of the National Institute, with the physicists and mathematicians being accorded the first class of that institution along with the largest number of sections (10). The second class was occupied by the moral and political sciences, while the third was the province of literature and fine arts. Napoleon was later to put his stamp on the institution in 1803 by eliminating the section devoted to moral and political sciences as being politically suspect and replacing it with two new classes: language and literature of France and ancient languages and history. Thus was Colbert's late seventeenth-century goal of uniting all of France's intellectual life under one well-supervised roof finally achieved.

Before the Academy was closed down in 1793, it had a role in persuading the Revolutionary government to continue the pattern of Pacific exploration that had been a feature of the old regime – one of the many ways in which the Revolution carried on aspects of the old order. When La Pérouse and his two ships failed to return from the voyage which set off in 1785, it was largely his scientific allies (and particularly members of the Society of Natural History) who brought about the voyage of Joseph-Antoine Bruny D'Entrecasteaux from 1791 to 1794. The primary (and predictably unsuccessful) goal of this expedition was to rescue La Pérouse and his men, as well as to strengthen France's presence in what was, from a European perspective, the largely uncharted Pacific. So akin were the two voyages that the scientific instructions given to D'Entrecasteaux were an adaptation of those given to La Pérouse under the signature of Condorcet as perpetual secretary of the Academy.[3] But despite such continuities with the old regime, at least some of D'Entrecasteaux's deeply divided crew absorbed the voyage into the narrative of the

---

[2] Hahn, *Academy*, p. 70   [3] Gascoigne, *Encountering*, p. 422

Revolution. For Louis Deschamps, a naturalist on the D'Entrecasteaux expedition, the voyage was an instance of the way in which France had the 'good faith to promote philosophy [science], that mortal enemy of superstition'.[4]

Later, when Napoleon had become First Consul, France renewed its claims to the Pacific and, in particular, to parts of Australia that had not been settled by the British, with the expedition of Nicolas Baudin from 1800 to 1804. Reflecting the high status science enjoyed under Napoleon, this expedition was the result of close consultation with major figures in the National Institute and the Museum of Natural History. On board were twenty-two scientists who outnumbered the senior officers, the source of much dissent.[5] For Baudin, the voyage was intended 'to destroy the prejudice of those who . . . fear the progress of reason and the arts'[6] – further testimony to the consonance between the Revolution and the values of reason and science.

The great institutional beneficiary of such voyages was the National Museum of Natural History, which, during the Revolution, was expanded and given greater resources. Indeed, it was granted considerable autonomy, with its support being seen as a natural part of the state's responsibilities, while the scientists were granted considerable freedom in their choice of research topics.[7] The nature of such research was, however, limited by the availability of specimens, which, in turn, was often dependent on state initiatives. Its considerable achievements, which included the work of Georges Cuvier on comparative anatomy and palaeontology, and Jean-Baptiste Lamarck's evolutionary theorising, were a source of national prestige. This helps to account for the alliance between the Museum and the state to obtain specimens through the imperial reach of the latter. Along with the state-supported voyages, specimens were confiscated from countries which came under the Revolutionary armies' control. Such botanical and zoological expansionism was consistent with the global role which André Thouin, the foundation professor of horticulture, envisaged for the Museum as a place which had become 'the central point for the reunion of the plants dispersed in the different parts of the world'.[8] One of the arguments advanced to the minister of the navy by Antoine de Jussieu, professor of botany at the Museum, for mounting the Baudin expedition was that it would help to make that institution 'more important than any other foreign establishment of the same kind'.[9] Such a reputation for pre-eminence was, Jussieu

[4] Gascoigne, *Encountering*, p. 421   [5] Gascoigne, *Encountering*, p. 421
[6] Gascoigne, *Encountering*, p. 422   [7] Mary Windsor, 'Museums', p. 53
[8] Spary, *Utopia's garden*, p. 92   [9] Gascoigne, *Encountering*, p. 423

added, also likely to be promoted by Revolutionary France's military successes, which brought in their wake confiscated specimens from other countries. Revolutionary France was, then, prepared to invest in the prestige accorded to the increasingly illustrious National Museum of Natural History, an outstanding example of the association between Revolutionary ideals and scientific practice. The patronage of Napoleon, a former scientifically trained artillery officer, of the Baudin voyage was consistent, too, with his role as a patron of learning. When, for example, he completed his conquest of Egypt, he established in 1798 a virtual branch of the National Institute, the Institute of Egypt, dedicated to the comprehensive study of its scientific and historical heritage.

The D'Entrecasteaux and Baudin voyages were, in part, a continuation of the Pacific voyages undertaken under the old regime by figures such as Bougainville and La Pérouse. But the Revolution had brought with it a break with royal patronage and the reconceptualisation of the major scientific institutions, such as the National Institute and the Museum, as being the expression of the will of the sovereign people. Voyagers, then, had to adjust to a different climate of opinion and rhetoric in order to gain the support of a new type of politics – one expression of which was greater centralisation of institutions such as the Institute and the Museum, and firmer control by the state.[10] Such state scrutiny brought with it greater demands for scientists to produce results from which the nation as a whole could benefit. The investment of the state in science owed much to the contribution that scientists made to Revolutionary victory, with the engineer and mathematician, Lazare Carnot, being entitled the 'organiser of victory'. Though Antoine Lavoisier was executed in 1794, as a former tax farmer he had contributed to the war effort through devising methods for the artificial production of saltpetre (needed for making gun powder) after the traditional source of supply from India was blocked by the British navy. The demand for metal in a suitable form for turning into canons by melting church bells was also met by French scientists. Overall, the experience of linking science and war helped persuade the Revolutionary state that science was useful.[11]

The consequent professionalisation of science[12] as scientists gained recognised social positions was also reflected in the growing interest of the Revolution in scientific education. Part of the Revolution's reshaping of national life the better to reflect the dominance of the nation state rather than the monarch was the transformation of old-regime royalist institutions into those which better reflected the ethos of service to the

[10] Harrison, 'Projections', p. 52   [11] Crosland, *Science under control*, pp. 20–21
[12] Gillispie, 'Science and politics'

state. The old Royal Engineering School at Mézières[13] and the School of Bridges and Highways helped to provide the foundation for a quite new institution, the École Polytechnique. This was founded in 1794, largely thanks to the endeavours of Lazare Carnot, with a title which emphasised training for his own profession of engineering: Central School for Public Works. A year later, however, it adopted its present title, which conveyed the way in which it provided a fundamental (and highly rigorous) mathematical and scientific education that could be then built on by specialised training in a range of fields at the Écoles d'application (Applied Schools).[14] It became a centre not only of elite scientific training, but also of major research in the area of the mathematical sciences. Reflecting the trajectory of the Revolution, in 1805 it was transformed by the then Emperor Napoleon into a military academy, with the students wearing uniforms and living in barracks. This was indicative of the way in which, during the Napoleonic period, science became more allied with the armed forces and more a part of a military state. This association continued to be a feature of the French state, limiting the contribution that science made to productive fields such as agriculture and industry.[15]

Along with the École Polytechnique, the Revolution gave birth to other institutions that promoted science. The École normale supérieure (Superior Normal School), intended to train elite teachers, was founded in 1795, and included an exposure to science. The third major educational institution founded in the mid-1790s was the Le Conservatoire national des arts et métiers (the National Conservatorium for the Arts and Trades) (founded 1794), to promote national productivity by cultivating engineering and the trades. This was in the Enlightenment spirit of the great Encyclopaedia of Diderot and D'Alembert, with its Baconian veneration for the trades. Rivalry with the English merchant and Royal navies prompted the establishment of a Bureau des longitudes in 1795. Its differences from its English prototype, the Board of Longitude, were telling. While the latter largely relied on individual initiative with rewards from the Board, the French set up a research institution.[16] This helps to explain why the French Bureau survives today as part of the state's promotion of science, while the Board of Longitude was abolished in 1828, since it was considered to have achieved the end for which it was founded. The Bureau had its base in the Paris Observatory and was to be particularly prominent in the field of astronomy.[17]

---

[13] Langines, *Conserving*, p. 259   [14] Fischer and Lundgreen, 'Recruitment', p. 553
[15] Williams, 'Science, education', *Isis*, pp. 380–1   [16] Schiavon, 'The bureau', p. 70
[17] Pyenson and Pyenson, *Servants*, p. 298

While these institutions were freshly founded during the Revolution, others (such as the Museum of Natural History) were refashioned from their old regime forms. Thus, the Collège royal (founded 1530) was rechristened the Collège national (from 1870 the Collège de France) and expanded to include a greater range of subjects (including science) than had been included in its foundation as a centre for the study of humanism. Thus, by 1798 its professoriate had expanded from the founding six to twenty.[18] One major set of institutions that had some difficulty in being accommodated within the Revolution's centralising template was the universities. These were suspect as being largely clerically dominated, and in their old regime form privileged corporations not directly under the control of the state. Accordingly, they were abolished in 1791, until the foundation of the Imperial University in 1808,[19] which brought all the university faculties throughout France under the control of one centralised body. This legacy of state suspicion of university autonomy was to have a profound effect on French scientific and intellectual life, with the universities playing less of a role than in other comparable societies. The inner elite were largely educated outside the universities, in many cases in institutions founded by the Revolution (or through the refounding of old ones). This meant greater dissemination of science among the French elite and a greater bonding of science with the instrumentalities of government.[20] Indeed, throughout the nineteenth century, most scientists were to be employees of the state.[21] In contrast to England, in the nineteenth century many of France's politicians and chief bureaucrats had received a scientific education,[22] providing the basis for a natural alliance between the state and science.

One way in which the spirit of scientific precision and rationality entered into everyday life was in the introduction of the metric system. This had been talked about in the old regime, but it had taken the shock of the Revolution to make such a change from traditional ways. Local systems of measurement were linked with local custom and identity, and the imposition of a single system for the whole of France would not have been politically possible in the old regime. Old regime work on a new system of measurement was carried into the early years of the Revolution, and in 1791 the Academy recommended the adoption of a unit of measurement known as the *metre*. This was defined as one ten millionth of the distance from the North Pole to the equator. Basing the measurement on nature was regarded as giving the new system a natural authority and

[18] Zeldin, 'Higher education', p. 77   [19] Outram, 'Politics and vocation', p. 29
[20] Alder, *Engineering*, p. 292   [21] Fox, *The savant*, p. 278
[22] Crosland, *Science under control*, p. 317

rationality as against the artificial systems hitherto in use. It was hoped, too, that such a definition might provide the basis for a universal system of measurement which could be used by all nations. Nationalism, however, was inscribed in the form of measurement, since the semi-meridian of longitude chosen as the basis was the one which went through Paris. Measuring this particular meridian took some time and was not complete when the then ubiquitous Lavoisier pushed the plan through the Revolutionary Convention in 1793. His motive was in part to delay the closure of the Academy, but his efforts were in vain, for this occurred a week later.[23] Given the premature nature of this adoption of the new measurements together with the distractions occasioned by the outbreak of war and the Terror, it was not until 1799 that the new system was meaningfully adopted. By then it was possible to draw on the accurate measurement of the semi-meridian which had been completed a year earlier. Together with the metre as the basis of length, the gram was made the basic unit of weight, being defined as the mass of one cubic centimetre of water.

Such new measurements, justified on scientific grounds, were a way of alerting the French population to the extent of the state's powers to break with the past. In practice, the old ways often continued, and the metric system was not made fully mandatory until 1837. Even more of a break with the old regime was marked by the Revolutionary calendar dating from the formation of the French republic in 1792. Its ten-day week and subdivisions of the year replacing the month were intended to break the traditional attachment to religious festivals and other traditional obser-vances. It was introduced by the Revolutionary Convention in 1793 at the same time as the adoption of the metric system. Another innovation adopted at this period of the Terror, which sought to completely remould French society, was a decimal system of time keeping. This, however, proved difficult to use, and was finally completely abolished in 1806, the same year as the abolition of the Revolutionary calendar. Ever the politi-cian, Napoleon, who was then in power, appreciated the limits of abstract rationality as a guide to the way in which the state interacted with the French people. It was Napoleon, too, who effectively suspended the metric system in 1812.[24]

As its architects intended, the use of the metric system became a badge of modernity and was adopted outside France in that spirit. It spread quite rapidly throughout Europe (with Portugal the first non-French state to adopt it), except to places like Russia that were dubious about the political consequences of going metric. Britain, too, resisted modelling its

---

[23] Alder, 'A revolution', p. 51   [24] Alder, 'A revolution', p. 42

weights and measures on those of its French foe. It also spread beyond
Europe to Latin America (with Chile the first to adopt it in 1848), and to
rapidly modernising Meiji Japan. All along, the French architects of the
metric system had hoped that their system based on nature and incorporating a spirit of system and order would become a universal one. When
the issue first arose in 1789, the opening year of the Revolution, the
dextrous statesman, Charles de Tallyrand, proposed involving the
British in discussions 'so that all nations might adopt it'.[25] What was
probably the first international science congress was held in Paris in
1798–9 to discuss the implementation of the new system of
measurements.[26] The metric system, then, was both devised for France
and its needs, but was also available to any state which wished to adopt it.
Such an ecumenical spirit reflected the belief of the Revolutionary leaders
that the scientific spirit which had shaped the metric system embodied
universal values that transcended any one state.

The same issue of the extent to which science could be both national
and cosmopolitan was a perennial tension within the scientific world.
Science, by its nature, is international, with specialists in one field readily
relating to the work of fellow specialists whatever their nationality. On the
other hand, with few exceptions, science takes place in a national setting,
often with the support of the state. Something of this dilemma was played
out in Revolutionary France, in the debates about how far the traditional
allegiance of the eighteenth-century academies to the larger Republic of
Letters could be maintained in the conditions of war. To a greater degree
than the Academy under the old regime, the major scientific institutions
of the Revolutionary period had become an integral part of the French
state, which brought with it the competitive spirit endemic to relations
between states. One solution that occurred to Napoleon was to make
prominent non-French scientists honorary Frenchmen. Hence, he
saluted the achievements of the Milanese astronomer, Bartholomeo
Oriani, by affirming that 'all men of genius, all those who have attained
a distinguished rank in the republic of Letters, are French, in whatever
country they happen to have been born'.[27]

Such an ingenious solution was not likely to work with countries, like
Britain, at war with France. There, the President of the Royal Society,
Joseph Banks, endeavoured to keep open some communication with the
National Institute of which he was made a member in 1801. He estab-
lished his allegiance to the Republic of Letters by such actions as endea-
vouring to use his influence with the British government to free the

[25] Heilbron, 'The measure', p. 219   [26] Crosland, 'The congress'
[27] Daston, 'Nationalism', p. 95

French geologist, Déodat de Dolomieu, who had been imprisoned by the Kingdom of the Two Sicilies, then an ally of Britain. With the increasing dominance of Napoleon, however, relations between states and their scientific establishments became ever more competitive.[28] Regulations introduced by him in 1807 largely subverted Banks's attempts to maintain contact – the modern state had proved more powerful than the Republic of Letters.[29]

Restoration France

The state structure devised by Napoleon proved remarkably long-lived, despite all the upheavals which followed his overthrow in 1815. The Bourbons returned with the ambition of reviving the old regime, but this had limited consequences for the pursuit of science. In 1816, the scientific first class of the National Institute was rechristened the Royal Academy of Sciences, though it remained within the overarching structure of the National Institute with its coverage of French intellectual life more generally. (The adjective 'Royal' was later removed after the Revolution of 1848 and the founding of the Second Republic). The name may have been changed to reflect a state once more focussed on the person of the king, but the substance of the Academy's activities changed little. One member was expelled for his Napoleonic loyalties, and some of the Napoleonic aristocracy were incorporated into its ranks, but this did not bring with it a major restructuring of the institution – especially as these new members from the aristocracy were at least meant to be ennobled on meritocratic grounds. As was so often the case, the scientists were willing to work with whatever regime provided them with the resources and recognition that enabled them to get on with their work. The consequence was that, as Maurice Crosland writes, the Academy became 'a significantly Royalist institution'.[30]

The transfer of loyalty from the Napoleonic regime to that of the Bourbons and the later changes in form of the French state were softened by a form of patriotic loyalty to the abstract entity of the French state. This was seen as transcending the different constitutional forms the state took over the course of the nineteenth century.[31] Such an outlook provided the basis for the further integration of science into the bureaucracy of the state[32] that had been in progress since the Revolution and even the old regime. Problems with pollution, which had led to an environmentally

---

[28] Daston, 'Nationalism', p. 104   [29] Gascoigne, *Science in the service*, p. 157
[30] Crosland, *Science under control*, p. 314   [31] Crosland, *Science under control*, p. 315
[32] Fox, *Savant and the state*, p. 2

innovative decree of 1810, continued to be monitored by leading chemists[33] – an instance of the continuity between the Napoleonic and the Restoration regimes. Members of the Academy were consulted on the scientific ramifications of the organs of state concerned with the promotion of industry and commerce. Lighthouses, for example, were overseen by a commission of the Ministry of Commerce on which the academicians were represented.[34] There was continuity, too, in the procedures of the Academy that had promoted the professionalisation of science. In particular, the insistence on publication continued, which advanced the discipline by focussing closely on one specialised area.[35] Though the Academy remained the dominant scientific institution, the diffusion of science among the elite was also assisted by the foundation of the Congrès Scientifique de France in 1833. This attempted to engage those with scientific interests in the provinces by moving yearly from city to city and drawing on local scientific societies for its membership.[36]

The Bourbon regime was committed to restoring the Catholic Church to something of its pre-eminent old regime position, but this did not cause great conflict with the Academy – even though an alliance was to develop between science and anticlericalism later in the century. Science appeared too remote from the central concerns of the Bourbons to be regarded with anything but neutrality, or even indifference.[37] One area where the Bourbons did provide positive support for a scientific venture was in the continuation of the tradition of French scientifically-informed exploration of the Pacific. From 1817 to 1840, the French sponsored eleven Pacific voyages, while the British dispatched only five.[38] The motives for sending them were the familiar mix of imperialism and scientific curiosity. Such motives were perhaps given particular impetus by a determination to show that a defeated France still retained pre-eminence in the area of science. Similarly, post-World War I Germany was later also to look to science to restore its international prestige, an instance of the way that science can act as one source of national pride.

Plans for Pacific colonies, however, gave way to expansion into more immediate territories in North Africa. The tradition of combining exploration with a scientific possession of the land continued. In the spirit of Napoleon's Egyptian institute, or the detailed scientific accounts of the Pacific, in 1837 the minister of war wrote to the Academy to draw up instructions for a voyage to Algeria. Such a voyage was intended to give a scientific claim on the land and to provide an overview of its resources.

[33] Le Roux, 'Chemistry', pp. 195–222.   [34] Crosland, *Science under control*, p. 404
[35] Crosland, *Science under control*, p. 30   [36] Bruce, *The launching*, p. 9
[37] Fox, 'Scientific enterprise'   [38] Gascoigne, 'From science to religion', p. 109

The result was the compendious thirty-volume *Exploration scientifique de l'Algerie* (Paris, 1844–67).[39] The specimens brought back by such voyages accumulated profitably in the Natural History Museum, helping to maintain the high reputation it had achieved in the Revolutionary and Napoleonic period.[40] Scientific research also benefited from the data on voyages accumulated by bureaucratic institutions of the state, such as the Ministry of Foreign Affairs or the navy[41] – an instance of the natural alliance between science and a state increasingly concerned with accumulating data.

Data became ever more important for many European states as the nineteenth century progressed, and particularly for France with its complex bureaucratic structures. French administrators were increasingly conscious of the need to govern not simply a state, but a nation which needed to be understood and, in particular, measured. Basic demographic information became the property of the nation as a whole rather than state secrets as it had been before the Revolution. This was consistent with a view of the state as a rational construction based on empirical data rather than the sanction of tradition.[42] Population figures for France had been based on estimates calculated in ever more sophisticated ways, culminating in the use of the calculus of probabilities by Laplace.[43] It was a mark of the extent to which the French state was increasingly ready to invest in data that in 1833 it re-established the Statistique Générale, which oversaw a full census every five years with the basic unit of measurement being families and households.[44]

German-Speaking Lands

France, then, emerged from the Revolution with an enhanced scientific reputation, largely thanks to its successful alliance with the state in the various incarnations that national government took in the first half of the nineteenth century. As the century progressed, France was increasingly mindful, however, that much of its scientific lead over other countries was being eclipsed. Of particular concern were the German-speaking lands, which hitherto had been overshadowed by the French scientific establishment. Where France had emphasised centralisation of its scientific institutions, the German principalities were characterised by a high degree of decentralisation. This followed naturally from the emphasis that Germany placed on the universities, a relatively neglected part of

[39] Crosland, *Science under control*, p. 391
[40] De Sauvigny, 'Science et politique', pp. 284–5
[41] Crosland, *Science under control*, p. 321   [42] Yeo, 'Social surveys', p. 86
[43] Rusnock, 'Quantification', p. 32   [44] Yeo, 'Social surveys', p. 87

French intellectual life. These were necessarily spread throughout Germany, since each German principality sought to maintain a university as part of its local identity; the close linkage between universities and the different princely states also reduced the tendency to develop a hierarchy of universities. This wide diffusion of the intellectual resources of the German lands continued to shape German scientific life even after the unification of Germany in 1871. The fact that Wilhemite Germany retained elements of a federal constitution also worked against overdue centralisation.

Much of Germany had been under French control during the Napoleonic Wars, leading to a reaction against things French and an increasing emphasis on what was distinctively German. One such institution was the university – at least when compared with France. Universities, then, received particular attention in the reforms of the German states prompted by French dominance. When the University of Berlin was founded in 1811, its architect, Wilhelm von Humboldt, entrenched a strong emphasis on the importance of research that influenced German universities more generally. Advancement within the university system was increasingly linked to research,[45] in contrast to the French universities, where the professor was first and foremost a teacher. As late as the 1930s, a French minister of education remarked that 'research is an irregularity toward which we turn a blind eye'.[46] The Grandes Écoles (such as the École Polytechnique) were also primarily teaching institutions that did little to introduce students to the conduct of research though providing a rigorous scientific education.

Science in Germany, then, was largely located in the universities, with less direct linkage with the central state than in France. Even the Berlin academy, which, tellingly, was refounded in 1812 with a German name – Akademie der Wissenshaften (Academy of the Sciences) – was effectively subsumed within the University of Berlin and lost its distinctive identity as an academy.[47] But, of course, in the German lands there was no central state until 1871, and the connections between science and the state took place at the level of the principality. Thus, maintaining the regional university and giving it scope to conduct research was made possible by the individual state governments. While in the eighteenth century universities were expected to be largely self-supporting, in the nineteenth the local principalities were more willing to provide support. This was prompted partly by competition with other principalities, but increasingly, from about the first third of the century, the local governments were

---

[45] Cahan, 'Institutional revolution', p. 6   [46] Gilpin, *France*, p. 97
[47] McClellan, *Science reorganised*, p. 255

willing to support forms of education that trained students for industry.[48] Germany, then, provided another form of state-science alliance made possible in part by its decentralised nature which allowed more direct access by professors to the apparatus of government. Such governmental expenditure helped make possible, for example, the celebrated chemical laboratory of Justus Liebig established at the University of Giessen (within the state of Hesse) in 1825.[49] This was to be a prototype of the laboratory in which German chemistry students were trained not only to imbibe the content of their discipline, but also how to conduct research. As always, the resources the state provided shaped the kind of science that could be practised, making chemistry, for example, a particular German strength.

Though the decentralised nature of German science had its advantages, there was evidently a felt need for some form of centralised exchange of scientific ideas, even if this was not possible at a government level. Hence, in 1822, twenty German scientists organised the Gesellschaft Deutscher Naturforscher und Ärzte (German Society of Nature Researchers and Doctors). This provided some scientific unity in the disunited German lands. Though open to members from all the German princely states, it avoided being too closely identified with any one principality by holding annual meetings in different cities. Membership was restricted to those who were actively publishing, but, nonetheless, within a decade the total number of members had risen to nearly 500.[50] It served as a model for the British Association for the Advancement of Science (1831) and the French Congrès Scientifique de France (1833), an indication of Germany's growing scientific influence.

Britain

For Britain, as for Germany, the French Revolution was a challenge to its established institutions and the ideological foundations on which they were based. The Revolutionaries claimed to be building a new form of state devised along rational lines, without any of the debris of the past. This challenged well-established rationales for the existing order in Britain based on tradition and religion. The Revolution, then, prompted a rallying to such symbols of the continuing order as the monarchy and the established Church of England. For some, the reaction had earlier been set in motion by the impact of the American Revolution, which, in

---

[48] Tuchman, *Science*, pp. 5, 17, 75   [49] Grove, *In defence*, p. 28
[50] Bruce, *The launching*, p. 8

a less threatening form, posed some of the same challenges as the French Revolution. All of this had its reflection in the career of the cause of political reform, which was kept at bay by a ruling oligarchy afraid of the spread of revolution until the passing of the Great Reform Bill in 1832. This mood of reaction shaped imperial as well as domestic policy – thus, Bayly characterises the period 1780–1830 as being one in which authoritarian governments were frequently installed in British colonies, with their power shored up by the traditional methods of pomp and circumstance to promote a hierarchical and even aristocratic order.[51]

This was not an atmosphere conducive to much innovation in the relations between science and the state, though the need for food supplies during the French wars did prompt some government support for agricultural improvement. Thus, the Board of Agriculture was founded in 1793, with an annual state grant of 3,000 pounds a year. It was hoped, however, that it would also be supported from private sources. Friction over government policy limited this, however, and it was dissolved in 1821.[52] Also closely related to the needs of agriculture was the Royal Institution, founded in 1799. It was established with a royal charter but was supported by private subscription, and its relations to the state were rather tenuous. In the period of the French Revolution, its safe political credentials were apparent in the extent to which it was dominated by landowners seeking agricultural improvement. After the Revolution, as Morris Berman has argued, it became more preoccupied with industrial improvements as its membership became more middle class in character. This did lead to it providing some advice to government on a range of subjects, including Michael Faraday's work on the maintenance of lighthouses.[53] Like the Royal Society, however, it was independent from the state, in conspicuous contrast to the way in which in France the major scientific institutions became part of the fabric of government.

Though resistant to change, the Royal Society remained the chief intermediary between science and the mechanisms of government during the period of the Revolution. Presided over by Joseph Banks until 1820, it reflected his love of the aristocratic and monarchical order with gentleman amateurs still playing a major part in its affairs. The rather loose structure of the Society had been magnified by the increase in the number of members from 125 in 1700 to close on 500 by 1800, with almost 60 per cent of these not having any scientific standing.[54] Reform did come in the wake of more general reforms to the constituted order in Church and state, of which the Great Reform Bill of 1832 was the

[51] Bayly, *Imperial meridian*, pp. 8–9   [52] Gascoigne, *Science in the service*, p. 128
[53] Berman, *Social change*, p. 173   [54] Gleason, *The Royal Society*, p. 3

dominant embodiment.[55] One such reform was the limiting in 1847 of the number of Fellows elected annually to only fifteen.[56] The overall effect of such reforms was to move the Society more in the direction of a meritocracy, with membership based on scientific eminence. As Roy MacLeod points out, however, this made the Royal Society both more rigorously scientific and more removed from more popular forms of science, including many forms of applied science and science education.[57]

The mood of reaction in the Revolutionary period did little to strengthen the credentials of science as an ideological defence of the constituted order in Church and state. Nonetheless, scientific concerns continued to infiltrate the mechanism of a government apparatus that had grown under the impact of war. The range of subjects on which Banks was consulted as President of the Royal Society conveys how far scientific issues were ineluctably becoming part of the province of government. In an effort to provide a more secure basis for taxation, the Excise Office commissioned Banks in 1791 to ascertain 'the just proportion of duty to be paid by any kind of spirituous liquor' – something which would require the expertise of those in the Royal Society versed in the use of instruments such as the hydrometer. More directly linked to the needs of war were the requests in 1801 to comment on 'the best covering for the floors of Powder Works', or 'Magazines to prevent the bad effect of friction', and that in 1803 to decide on 'the best mode of preserving Flannel for Cartridges' – something that was meant to draw on Banks's knowledge of entomology. In 1809, the Victualling Office, a branch of the navy, wanted his opinion on a newly invented iron water storage vessel for use at sea.[58]

The armed forces were one bridge between science and the state, the issuing of currency was another – one which had the important precedent of the appointment of Isaac Newton as Warden and then Master of the Mint from 1696 to 1727. Banks, then, served on the Privy Council Committee on Coins from 1787. This oversaw the recoinage of copper money in 1797, and benefited from scientific experiments on the wear of metal by two fellows of the Royal Society, Henry Cavendish and Charles Hatchett.[59] The growing sophistication of currency brought with it the need both in Britain and elsewhere for further scientific advice as when, in 1819, Banks was placed on a committee to enquire into methods of preventing the forgery of banknotes.

[55] Foote, 'The place'   [56] Hall, *All scientists now*, pp. 143–4
[57] MacLeod, 'Whigs and savants'   [58] Gascoigne, *Science in the service*, pp. 25–7
[59] Gascoigne, *Science in the service*, p. 121

Banks was also called upon for advice on voyages of exploration, which necessarily were limited during the Napoleonic wars given the shortage of shipping. As part of Banks's oversight of the Australian colonies, however, he persuaded the Admiralty to put in motion a circumnavigation of Australia from 1801 to 1803 under the command of Matthew Flinders. On board was the botanist Robert Brown, who, on return, became Banks's librarian. After Banks's death in 1820, his massive natural history collections passed to the British Museum, where, from 1827, Brown was 'Keeper of the Banksian Botanical Collection'. These collections were, in turn, absorbed into the Museum of Natural History founded in 1883. Similarly, Sir Hans Sloane's large natural history collections had passed to the British Museum in 1753, another instance of a characteristic British phenomenon: the absorption by an expanding state of a private initiative.

With the end of the Napoleonic Wars in 1815, the British state began to attend to some of the more pressing issues brought about by the changes to a society increasingly being reshaped by industrialisation and commerce. One of the most urgent such concerns was, as in France, to replace regional systems of measurement with standard measures for the country as a whole, and to reduce the ten different systems of measurement to a single one.[60] Instead of folk measurements based on tradition, such as 'the Winchester bushel', the British government sought standard measurements based on scientific criteria. Accordingly, in 1817–19 Banks was made chairman of a parliamentary committee to consider weights and measures, which eventually resulted in the state-approved system of imperial measures in 1824. Hostility to France was an obstacle to adopting the metric system with its more overt scientific foundation. British measures were, however, given a scientific definition with the use of the amplitude of a swinging pendulum.[61] This was in conspicuous contrast to the basis of the French system with its use of the semi-meridian passing through Paris.

The end of the war also led to more expeditions of scientific discovery, which owed much to the initiative of John Barrow, a secretary of the Admiralty, who, in 1815, became a member of the Council of the Royal Society and a strong advocate of Admiralty-supported voyages of exploration. The polar regions were a particular focus of exploration, with the voyage of David Buchan and John Franklin to the Arctic in 1818 being the first of a number of expeditions to that area. The most successful of these voyages was that of James Clark Ross to Antarctica from 1839 to 1843, in search of the southern magnetic pole.[62] On board as Assistant Surgeon

[60] Allen, *The institutional revolution*, p. 33   [61] Gascoigne, *Science in the service*, p. 28
[62] Hall, *All scientists now*, p. 209

was the botanist, Joseph Hooker, thus carrying on the tradition that had been established by Joseph Banks's presence on Cook's *Endeavour* voyage of major naval voyages taking with them a naturalist. In the same spirit the Admiralty facilitated the most epochal of all scientific voyages by allowing Charles Darwin's presence on the *Beagle* voyage of 1831–6. The pre-eminence of the Royal Navy provided a major link between science and the state, and played a role in shaping British science to take particular account of natural history as well as the astronomical and physical sciences that were generally the major scientific focus of such voyages.

The end of the Napoleonic Wars in 1815 and the coming of the age of reform from 1832 also brought with it a willingness on the part of the state to engage more directly with the major scientific institutions. In 1820, lobbying from both the Royal Society and the Board of Longitude led to the government establishing an observatory in Cape Town. Relations with the Royal Society were strengthened in 1825 with the creation by the monarch of two royal medals to be bestowed by the Society on two fellows who had published the most important discovery in the previous year. The importance of prestige as a motive for acting as a patron of science was apparent in the prime minister's letter to George IV, urging the endowment of such medals as a way of enhancing royal glory.[63] What the Society really needed, however, was finance for promoting scientific research – which accounts for the decision in 1900 not to found more medals.[64] The beginnings of such direct funding for science using the Royal Society as the intermediary came in 1849 with the establishment of the government grant. Though at first a modest amount, it created the beginnings of a quite new relationship between science and the British state, with government funding being made available for the promotion of science without direct oversight by the state. It also meant a new role for the Royal Society as a patron of science. So much of a break from tradition was this that in 1855 the grant was abolished; it was, however, reinstated shortly afterwards and made larger in 1876.[65]

Science, then, was increasingly being recognised as a concern of the state, the governmental apparatus of which was steadily expanding throughout the nineteenth century. One index of such an expansion was that the number of governmental officials in 1827 was some 40 per cent higher than in 1797.[66] The growing range of state activities brought with it new governmental alliances that somewhat weakened the Royal Society's almost exclusive role as scientific adviser. Kew Gardens in 1840 was placed directly under state control, having been previously

[63] Gleason, *The Royal Society*, p. 93   [64] MacLeod, 'Of medals and men', p. 100
[65] Hall, *All scientists now*, pp. 163–4   [66] Gascoigne, *Science in the service*, p. 201

been in a more marginal position as a royal garden.[67] This change reflected its growing importance as the centre of a worldwide network of botanical gardens intended to facilitate the transfer of plants around the world, the better to serve the British economy. After the rechristening of the Astronomical Society of London (founded 1820) as the Royal Astronomical Society in 1830, it was granted a role hitherto an exclusive Royal Society preserve: that of acting as Visitors of the Royal Observatory at Greenwich.[68]

One long-standing scientific body was abolished with the shutting down of the Board of Longitude in 1828, on the grounds that it was superfluous now that methods of determining longitude had been discovered. In its place was established another body to carry on some of its functions, the Resident Committee for Scientific Advice for the Admiralty, with three advisers who included Michael Faraday of the Royal Institution. This was short-lived, being abolished in 1831, but soon was replaced by a body that linked science more closely with that central state institution, the British navy: the Admiralty Scientific Branch founded in 1831. Among its responsibilities was overseeing the observatories at Greenwich and Cape Town. The scientific projects of the Admiralty included the standardisation of ships' logs in such a way that they could be used for meteorological research, another instance of the developing ties between the armed forces and scientific research.[69] The Admiralty also participated in the study of tides, providing much of the empirical evidence for the theoretical studies by William Whewell. The outcome was the issuing of tide tables that by mid-century provided information on over one hundred ports within Britain and overseas.[70] The armed forces and the Admiralty, in particular, provided a natural link between scientific activity and an expanding state.

One of the most direct ways in which the state drew on scientific expertise was in the mapping of the land. The Ordnance Survey office had been founded in 1791 to achieve such a goal, but the project grew in size to the point where a new government agency, the Geological Survey, was established in 1845 to deal with that particular specialised branch of mapping.[71] The Geological Survey, in turn, was to prompt the foundation of the Government School of Mines and Science Applied to the Arts in 1851. Chemistry also received government recognition with the establishment of the laboratory of the government chemist in 1843, and the foundation of the Royal College of Chemistry in 1845.[72] But the tradition

---

[67] Brockway, *Science and colonial expansion*, p. 169   [68] Hall, *All scientists now*, p. 183
[69] Naylor, 'Log books', p. 771   [70] Reidy, *Tides*, p. 234   [71] Close, *Early years*, p. 62
[72] Cardwell, *The organisation*, pp. 66–7 and Gummett, *Scientists*, p. 125

of relying on private initiative when possible and only involving the state when necessary still continued. The Rothamsted Experimental Station founded in 1843 by the owner of a lucrative fertiliser business was a major centre of agricultural research and remained self-funded until 1909.[73]

Along with the growth of such instrumentalities that linked science and the apparatus of government, the period of reform saw the growth of an alternative association to promote the study of science: the British Association for the Advancement of Science (founded 1831). This was not intended to supplant the Royal Society, but rather to complement it by bringing scientific discussion and education to a much wider public than the increasingly research-focussed Royal Society.[74] Hence, it was open to any member of a scientific society in the empire. Its growing size made it an influential lobby group – one of its goals being to persuade the state to support more scientific research. It also undertook some research itself, founding in 1842 what it termed a physical observatory at the defunct royal astronomical observatory at Kew.[75]

The growing interest in applying science to a range of phenomena extended to approaching social problems using scientific techniques. One reflection of this was the way in which a statistical section of the British Association for the Advancement of Science was founded in 1833, in part as a result of the initiative of the pioneer demographer, Thomas Malthus. A year later, he was involved in the founding of the Statistical Society of London.[76] This growing interest in what William Petty had termed in the late seventeenth century 'political arithmetic' had earlier been evident in the way in which Thomas Malthus gave lectures at the East India Company training college at Haileybury in 1805 on 'the branch of the science of a statesman or legislator'.[77] In exposing future colonial administrators (and chiefly those in India) to such an approach to government, Malthus was helping to implant the view that statistical methods offered a key to orderly rule both at home and abroad. Among the results of such a mentality were the categorisation of the Indian population according to religion, and the reduction of the many gradations of the Hindu caste system into a few statistically manageable categories.[78]

At the level of the state the growing sway of political arithmetic is most evident in the foundation of the census in 1801 – a device which enabled government to calculate its human resources just as the Ordnance Survey

[73] Agar, *Twentieth century*, p. 51; Knight, *Nature of science*, pp. 151–2
[74] Hall, *All scientists now*, pp. 327, 337
[75] MacDonald, 'Making Kew Observatory', p. 432
[76] After 1887, the Royal Statistical Society   [77] Winch, 'The science', pp. 83, 63
[78] Yeo, 'Social surveys', p. 30

and Geological Survey enabled it to calculate its physical resources. As was so often true, this innovation was the fruit of war, with the census being justified in part on the grounds that it would enable the calculation of the available manpower to fight Napoleon. Another consideration was calculating the amount of food needed to feed the population at a time of poor harvests and enemy blockades. John Rickman, who organised the first census, made clear in a letter to Joseph Banks the connection he drew between the census data and firm government control. 'The favourite object of my life', he wrote in 1805, 'is to distribute England into such orderly Divisions or Districts, that information may be obtained and good government enforced in the most effectual manner.'[79] The information that the census yielded to a government increasingly interested in rational planning prompted its repetition every ten years, until it was absorbed more professionally and securely into the mechanisms of the central state when the 1841 census was taken over by the General Register Office.[80]

Collectively, the Ordnance Survey, Geological Census and census contributed to the growth of Britain as a 'data state' with an increasing capacity to control its population and resources. From 1833, too, there was readily obtainable information on the pattern of trade with the foundation of the Board of Trade's statistical section.[81] The amassing of such data was a departure from earlier British traditions of not allowing the state too much information on its citizens, lest its power increase to unacceptable levels. The willingness of the British to tolerate a regular census was an index of the way in which the British state was expanding in size and complexity. Such growth was a consequence of fighting the French Revolutionary forces and having to contend with the social problems generated by the increasing sophistication of the economy in the age of the Industrial Revolution. The weight of tradition and centuries of opposition to royal absolutism, however, stood in the way of too great an expansion of state power. This was reflected in the state's role as patron of science. Where possible, the tradition of independent scientific activity was continued. Apart from a few small concessions, the Royal Society had to rely on its own resources, as did the Royal Institution and the British Association for the Advancement of Science. The statistical information that was becoming increasingly important was in part the product of private organisations, such as the statistical branch of the British Association for the Advancement of Science and the Statistical Society of London. State support for science was generally limited to areas where

[79] Gascoigne, 'Joseph Banks, mapping', pp. 165–6
[80] Higgs, 'The struggle'. I differ here on the importance of military mobilisation, which was cited in the parliamentary debates as a major reason for the census.
[81] Yeo, 'Social surveys', p. 87

there was an immediate application to the needs of government. Government support for scientific research on a larger scale was to await the emergence of a state that took a wider view of its remit and responsibilities.

The United States

As a state forged by revolution, the ideological justification of the USA's break with the British old regime was central to the nation's self-understanding. The Revolution's most articulate spokesman, Thomas Jefferson, linked the American Revolution with the canons of reason exemplified by science. The Declaration of Independence, the fruit of Jefferson's fertile pen, spoke of the justification embodied in 'the laws of nature and of nature's god'. Looking back on the course of the American Revolution, Jefferson again portrayed it as part of the advance of reason as embodied in science: its beginnings, he wrote, were part of a rebellion against 'monkish ignorance and superstition', a rebellion which embraced 'the rights of man' and the 'general spread of the light of science'.[82]

Yet the very readiness with which Jefferson could invoke the values of science and the Enlightenment were testimony to their diffusion before as well as after the Revolution. Science formed a part of polite culture in the British Thirteen Colonies and provided something of a foundation on which the new nation of the United States could build. The most conspicuous embodiment of pre-Revolutionary interest in the sciences was the American Philosophical Society, which was founded in Philadelphia in 1769 as a semi-public body. It was formally chartered by the Pennsylvania Assembly, and also received from it some limited funding both before and after the Revolution.[83] Its links with the workings of government were strengthened during the 1790s, when Congress met in Philadelphia while Washington was being built.[84] It remained, however, a voluntary body without any formal role in directing the nation's affairs, and its influence was largely restricted to Philadelphia. It was complemented rather than challenged by the founding of a new scientific society, the Boston-based American Academy of Arts and Sciences, in 1780. The American Academy's Revolutionary credentials were more evident than the American Philosophical Society being established in the midst of the War of Independence by the Massachusetts Assembly in 1780, with its early members including such notable revolutionaries as George Washington, Benjamin Franklin, John Adams and John Hancock. Both

---

[82] Price, 'The scientific estate', p. 88   [83] McClellan, *Science reorganised*, p. 141
[84] Dupree, *Science*, p. 18

bodies might claim in their titles to be American-wide bodies, but, in practice, their influence was largely regional. It was not until late in the nineteenth century that truly national scientific bodies were formed that drew on local networks formed by such early societies and the largely private universities.[85]

Translating the gentlemanly interest in science that prompted the foundation of these two societies into part of the structures of the state was, however, a slow and often contested process. From the early debates of the new United States there was opposition to any move to expand the powers of the federal government, and this was to include measures for the promotion of science. This accounts, for example, for the opposition to the foundation of a national university despite the advocacy of the iconic figure of George Washington. There were, however, some measures that could not be avoided if the federal government was to function. One such was the holding of a census to determine the size of federal electorates, a measure embodied in the constitution with the requirement that a census be held every ten years from 1790, to take account of changing levels of population.[86] It was an early step towards the building up of a 'data state', the government of which depended on the accumulation of information on the nature of the population and its needs.

What science the early United States government did promote was largely linked to the direct needs of government. The need for currency prompted the establishment of a mint, which, as in Britain, was seen as being an institution linked to science since it required a knowledge of metallurgy. Jefferson, then, appointed the astronomer David Ritterhouse (Benjamin Franklin's successor as president of the American Philosophical Society) as the founding director of the mint. This tradition of linking science and the mint continued until 1853, when, for the first time, a director was appointed without some scientific credentials.[87] The Mint provided a means of Jefferson institutionalising one of his pet reforms, the introduction of decimal currency. Jefferson was thwarted, however, in his attempt to introduce metric weights and measures, reforms which he saw as embodying a truly scientific spirit. Although the forces of tradition were too strong to abandon the old measures, Jefferson did give the republic's weights and measures a scientific foundation by reference to the pendulum measure[88] – as occurred in Britain.

During Jefferson's time as the third American president from 1801 to 1809 he did much to establish institutions that provided the foundation of future alliances between the state and science. These governmental

[85] Dupree, 'The national pattern', p. 31   [86] Yeo, 'Social surveys', p. 86
[87] Dupree, *Science*, pp. 17–18   [88] Dupree, *Science*, p. 18

instrumentalities rarely promoted disinterested research and were not intended to do so. Jefferson himself emphasised the importance of science contributing to practical outcomes. His work on weights and measures provided the foundation for the future Bureau of Standards, which began life in 1830 as the Office of Standard Weights and Measures within the US Department of the Treasury. Another particular interest of Jefferson was meteorology, and it was during his incumbency that a national weather bureau was established.[89]

One of the most significant institutions for the promotion of science was the Coast Survey, founded in 1807, while Jefferson was president. The act establishing it emphasised its practical advantages and the need 'to cause a survey to be taken of coasts of the United States, in which shall be designated the islands and shoals and places of anchorage'. Achieving such ends required a good deal of scientific expertise with the need for establishing points of reference near the coast by astronomical calculation, using these to build up networks of precise triangulation. Such calculations were then combined with topographical survey of the coast and hydrographic survey of the waters near it.[90] In the first half of the nineteenth century, then, the Coast Survey was to be one of the most prominent of government scientific instrumentalities, providing a seedbed for future research as well as routine charting. One example of its scientific expansionism was its study of magnetism, which it sought to justify on practical grounds.[91] Such straddling of practical tasks and scientific research was made a part of the culture of the Coast Survey by its highly influential superintendent from 1843 to 1867, Alexander Dalles Bache (grandson of Benjamin Franklin). Under his direction, naturalists were included in the crew of survey vessels and were encouraged to use what opportunities that presented themselves to pursue their interests.[92]

As always, military imperatives provided a bridge between science and the state; naturally, then, there were pressures for the Coast Survey to be absorbed into the navy, which occurred in 1834. Its time under naval control was, however, surprisingly brief, with civilian control being re-established in 1836. The navy was more directly involved in the creation of the United States' first national observatory, the Naval Observatory. It is an indication of the lack of congressional support for institutions perceived as being for the promotion of scientific research that this emerged from the navy, being justified on the grounds of promoting accurate navigation and time keeping. Its lowly place in the institutional

[89] Bedini, *Thomas Jefferson*, pp. 448, 491–2
[90] Wraight and Roberts, *The Coast and Geodetic Survey*, p. 5   [91] Dupree, *Science*, p. 102
[92] Benson, 'Field stations', p. 82

hierarchy was underlined by its foundation in 1830 as part of the Depot of Charts and Instruments. As with many government instrumentalities impinging on science, it was, however, to grow in stature, and in 1842 was made a national observatory. Along with the navy, the army also provided some possibilities for linking science with government. In 1802, a year after Jefferson became president, the Army Corps of Engineers was established along with the West Point Military Academy. The two were related, since in its founding years one of the major functions of West Point was to produce both military and civil engineers. In this it resembled the École Polytechnique, and the resemblance became more marked under the Academy's influential director, Sylvanus Thayer (1817–33), who modelled much of the curriculum on its French counterpart.[93] As with the École Polytechnique, the highly mathematical curriculum provided the basis for some original research by faculty members, though teaching remained the primary function of the institution. West Point's example in providing the only technical education available up to the 1820s also prompted imitation by other institutions, including Harvard and Yale.[94]

Along with the needs of war, one major concern of the early republic, which drew on scientific expertise, was the exploration of its territory. In this respect, the United States resembled Russia as a nation with a huge hinterland that required systematic mapping to be exploited. Famously, it was under Jefferson that the Lewis and Clark expedition (1804–6) was commissioned. Intended to explore Jefferson's recent Louisiana purchase and to map a path to the Pacific, the expedition also involved scientific surveys of the natural history of the area, including its human population. As part of the preparations for the expedition Lewis and Clark took advice from scientists on how, for example, to use scientific instruments to conduct astronomical observations, along with the identification of potentially useful minerals.[95] Numerous other expeditions by both land and sea followed which, again, combined routine mapping with the recording of scientific data. One stimulus for exploration was to establish a route for a transcontinental railway, the mission of some eighteen exploring expeditions led by the Army Corps of Engineers from 1848 to 1850.[96] Another was to establish clear boundaries with Mexico and Canada, a state-defining enterprise in which the Army Corps of Topographical Engineers was again active.[97] Exploration extended to surveying lands adjacent to the United States or which were important

[93] Dupree, *Science*, p. 36   [94] Sinclair, 'The promise', pp. 250, 161
[95] Reidy, *Exploration and science*, p. 140   [96] Reidy, *Exploration and science*, p. 145
[97] MacLeod, 'Discovery and exploration', pp. 51–2

for US commerce. American whaling interests in the Pacific were a stimulus for the Charles Wilkes expedition of 1838–42 that traversed much of the Pacific, as far as Antarctica. It was an expedition in the tradition of the great eighteenth-century European voyages of discovery, with close attention to possible scientific opportunities. It marked, then, the growing self-confidence of the United States and its determination to take its place in the roll call of exploring powers.

The huge collections of objects gathered by such expeditions in many cases made their way to the Smithsonian Institution, founded in Washington in 1846. This was the nearest thing to a governmentally-funded research centre founded by the federal government in the period up to the Civil War. Its origins derived from an unexpected bequest which, after much rumination by Congress, led to a museum that was also intended to act as a research centre. It was here that many of the larger scientific implications of the expeditions of the 1840s and 1850s, and the collections they amassed, were made known through published versions of those expeditions.[98] Such activity constituted a form of nation building, as it bolstered the central government's claim to the larger territory and resources of the United States, which had been augmented enormously by Western expansion and annexation of territory after the Mexican War of 1846 to 1848.

The state's patronage of science in the period from the Revolution to the mid-century was, then, largely utilitarian. It was chiefly linked to such state-defining activities as establishing coinage and weights and measures, determining the nation's land and sea borders through the Coast Survey's mapping of the Atlantic and Pacific coastlines and the Army Corps of Topographical Engineers' surveying of its land borders. Though science had formed a part of the Enlightenment-tinged ideology of the Revolution, the early republic was too busy establishing its foundations as a state to devote many resources to the pursuit of science for its own sake. In his *Democracy in America*, Alexis de Tocqueville (1835–40) commented that 'in America, the purely practical part of the sciences is admirably cultivated and great care is taken with the theoretical part immediately necessary to application . . . but there is practically no one in the United States who concerns himself with the essentially theoretical and abstract part of human knowledge'.[99] A partial exception to the neglect of science for its own sake was the establishment of the Smithsonian Institute, the result of a fortuitous bequest that prompted the United States' government to move in an unexpected direction. What the period 1776 to 1850 had done, however, was to translate the scientific strands of

Revolutionary ideology into ongoing institutions that shaped the values of the Republic. Appropriately, Jefferson was to be a major figure in both developing the justification for revolution and the drawing together of science and the early workings of government. It was on such foundations that the United States was later to build one of the most productive of all partnerships between science and the state.

The age of revolution left different inflections on the partnership between science and the state. In France, it helped to establish science as part of the governing ideology of the nation, as it did, in much more muted form, in the United States. On the other hand, countries like Britain and Germany resisted the influence of the French Revolution and, with it, the use of scientific ideology as a weapon against the old regime. All Western countries, however, faced similar problems in consolidating the position of the state, and these provided possibilities of an alliance between science and the state. The need for uniform measurements, reliable coinage and the mapping of borders all were a concern for states growing in size and complexity. So, too, was the need for accurate statistical information about the nation's population and, in particular, its size. This applied particularly to Britain in coming to terms with the repercussions of the Industrial Revolution. State funding for science, then, was generally linked to a definite practical outcome, though the institutions for the promotion or teaching of science founded or refounded by the French state in the Revolutionary period ensured its continuing reputation as the leading scientific nation. Its lead, however, was to be challenged by the German lands, where original research was supported through the subsidies granted by the governments of the princely states to the universities.

Though limited by the French wars, exploration by France and Britain continued. This offered another avenue for state support for science, combined with other goals such as imperial and strategic advancement. In a new state such as the United States, the state also sponsored a considerable amount of internal exploration with the goal, among others, of establishing clear borders. Given the number of regimes that existed in post-Revolutionary France and elsewhere in Europe, the period 1789–1850 provides an illustration of the extent to which scientists were (and are) willing to work with quite diverse forms of state. Different states might have shared similar problems, but their way of addressing these with the aid of science varied according to the different patterns of government. Different states evolved different scientific institutions and gave varying scope for the emergence of a distinct scientific community. Building bridges between science and the state was an incremental process with varying strategies and tactics on both sides.

6   An Expanding State, 1850–1914

When the state first emerged as a distinctive form of political organisation its functions were limited. Primarily, the state's tasks were to organise for war and to maintain internal law and order. Over time, its functions began to expand, and they were to do so considerably in the period from the mid-nineteenth century up to World War I. A number of developments help to explain the widening of the state's remit in this period. As the Industrial Revolution gathered pace, the consequent social problems – and particularly those posed by ever more crowded industrial cities – meant more involvement by the state in providing remedies. The spread of democratic ideas, even when held in check by traditional regimes, exerted pressure on governments to be more responsive to an increasingly numerous and literate population – if only to ward off those urging more radical solutions. The nature of industrialisation also began to change, moving from the relatively simple machinery of the first Industrial Revolution based around textiles and iron-working, to the scientifically more sophisticated industries of the second Industrial Revolution based around the chemical, electrical and metallurgical industries. In different ways, then, the state was becoming more of a patron of science, whether in providing services to promote the health and wellbeing of an expanding population, or to create the scientific capital that could help fuel an economy ever more dependent on knowledge.

Germany

In the integration of the state into the phenomenon which has been described as 'the capitalization of knowledge',[1] Germany was to take the lead. As it became evident that scientific knowledge could be valuable for industry, the state took an increasing interest in the institutions which produced such knowledge. As we saw in the previous chapter, the origins

[1] Shinn, 'The industry, research, and education nexus', p. 134

98

of this commitment to the advancement of a nexus between scientific research and industrial application lay in the revitalisation of the university system in the wake of the Napoleonic Wars. Under these Humboldtian reforms, the role of research was given a major role within the universities, in contrast to the almost exclusively teaching role that had previously prevailed. This research ethic gradually took deeper root within the universities, largely thanks to the willingness of state governments to finance research facilities in the expectation that these would provide economic returns. This involvement by state governments in the promotion of research was steadily gaining ground in the 1850s and 1860s before the unification of 1871 brought about further changes. One way of promoting research within the universities was the creation of research institutes – these remained a part of the university but were answerable directly to the state government, a further instance of the state being drawn into the task of financing scientific advance.[2] The small state of Baden, for example, spent what was then the enormous sum of 450,000 gulden on new scientific institutes, with the jewel in the crown being that at Heidelberg under the physician and physicist, Hermann von Helmholtz.[3]

Such research institutes were to grow in size and importance as state governments became ever more persuaded of the importance of scientific research as a means of economic growth and social improvement – in part because the state administrators were themselves university graduates and reflected academic values. It is a process epitomised by the changing role of physics institutes, which has been traced by David Cahan.[4] Physics research institutes had become a common feature of the German university by the 1860s, but they were relatively simple with limited space or facilities directly allocated to them. By 1914, all universities had modernised institutes with a modern collection of instruments, a laboratory and three or four assistants. Such institutes were intended for teaching as well as research, with space for student practical lessons – hence, graduates were trained in the techniques of research.

The traditional German university was not, however, always a congenial setting for applied research given the prestige accorded by the German professoriate to pure research. The result was that the German states invested in a new form of an institution of higher education, the Technische Hochshule (Technical Institutes). The universities looked with some dismay at the growth of these competitor institutions and were particularly jealous of the universities' traditional prerogative of granting

[2] McClelland, *State, university and society*, p. 286   [3] Tuchman, *Science*, p. 168
[4] Cahan, 'Institutional revolution'

degrees. Hence, the fact that the technical institutes were allowed to award doctorates after 1900 marked a significant victory for the new institutions. It was also an index of the support they received from the state which had come to view them as alternative sites of higher education particularly well suited to industrial research – by about 1880, for example, the Prussian technical institutes were granted greater recognition by being transferred from the control of the Ministry of Trade to that of the Ministry of Education.[5]

The expanding demands of the 'capitalization of knowledge' meant, then, the growth of a variety of institutions to link scientific research and training with industrial demand. In establishing such a pattern, the dye industry played a particularly important role as a leading sector[6] – thus helping to establish German pre-eminence in chemistry. Chemical graduates were overwhelmingly employed in the German chemical industries, underlining the increasing symbiosis between state-funded institutes of higher education and scientifically informed industry. State regulation could help foster such a symbiosis: the first patent law of the newly united Germany encouraged the transference of research to industry.[7] As the value of research became ever more evident to industry, some companies set up their own research institutes. In 1891, for example, the chemical and pharmaceutical company Bayer established a new laboratory in which the process of invention and research became routinized and controlled by a hierarchical industrial management.[8]

Industry may have begun to supplement the research conducted by the universities and technical institutes, but there were areas of science that were insufficiently directly profitable to entice investment. This meant increasing scientific reliance on the state as Germany evolved new institutions to provide a scientific foundation for the long-term promotion of industry and, with it, of science itself. The first such major new institution was the Physikalisch-Technische Reichsanstalt (Imperial Physical Technical Institute) in 1887. The adjective 'imperial' had particular meaning as an initiative of the central government of the newly united Germany rather than of the particular states. This catered particularly to the needs of the fast-moving electrical industry by providing appropriate standards. It also included facilities for more fundamental research in experimental physics.

The three figures who were most active in its foundation embodied the different functions which the Institute was expected to fulfil. Firstly,

[5] McClelland, *State, society, and university*, pp. 304, 306   [6] Gilpin, *France*, p. 93
[7] Meyer-Thurow, 'Industrialisation of invention', p. 368
[8] Meyer-Thurow, 'Industrialisation of invention', pp. 370, 379

Werner von Siemens, of the giant electrical company, Siemens, whose conception of the Institute was as a centre for promoting the long-term advancement of industries dependent on physics. Siemens considered the physical sciences particularly needed such support, since industry itself was less likely to conduct research in these areas requiring large-scale long-term funding than the more obviously industrially lucrative chemical industries.[9] The very active support which Siemens gave to the project indicated that support for its goals could come from industry, an assumption given substance by the way in which Siemens donated the land on which the institute was built just outside Berlin at Charlottenburg. A second moving force was the eminent physicist, Hermann Helmholtz, then at the University of Berlin. His appointment as the first director of the Institute gave it indisputable scientific status and a reputation for both pure and applied science. The third major figure was Wilhelm Foerster, president of the Prussian Normal-Eichungs Commission (Prussian Office of Weights and Measures) (founded 1816), an organisation which shared the Institute's goals of standardising measurements and promoting research in precision technology. As a prominent member of the Prussian bureaucracy, his presence may have helped ward off some of the resentment from the particular German states (and particularly Prussia) at the foundation of an imperial institute rather than one under state government. Siemens's role in raising money from industry, thereby creating a partnership between government and industry, also helped reconcile the separate states to the budget savings that followed from the imperial government acting as the patron of the enterprise.[10] State opposition would also have been lessened thanks to the support from the Crown Prince Frederick William, an invaluable connection in gaining support for a new institution that might be seen as venturing onto the territory of existing entities: much time, for example, was devoted to negotiation with the Academy of Sciences of Berlin.[11]

The successful foundation of the Institute marked a new model for government-industry relations outside the universities and other institutes of higher education. Its success prompted others to emulate this novel institution. Within Germany, other disciplines sought a similar fusion of research in both the pure and applied sciences, with the support of both government and industry. In 1905, an equivalent body for the biological sciences, the Kaiserliche Biologische Anstalt (Imperial Biological Institute), was created with active support from the central state in the form of the Reich Ministry of the Interior.[12] The chemical

[9] Beyerchen, 'On the stimulation', p. 148   [10] Szöllösi-Janze, 'Science', pp. 348–50
[11] Pfetsch, 'Scientific organisation', p. 575   [12] Szöllösi-Janze, 'Science', pp. 339–60

industry looked on at the workings of the physics institute with considerable interest and contemplated an imperial chemical institute, but this did
not come to pass until 1921 with the foundation of the Imperial Institute
of Chemistry.[13] Internationally, other countries responding to similar
developments as were occurring in Germany also drew on the physics
institute example as providing a new model for promoting scientific
research. Thus, the British National Physical Laboratory (1900), the
US Bureau of Standards (1901) and the French Laboratoire national d'essai
(1901) (National Experimental Laboratory) were founded to perform the
same roles as the Imperial Institute – the successful transplantation of the
model meant growing competition so that Germany's pre-eminence in
metrology was under challenge by 1914.[14]

The successful creation of a new model for the promotion of research
was a stimulus for Wilhelmite Germany to develop another. The aim of
this new institution, Kaiser Wilhelm Gesellschaft zur Föderung der
Wissenschaft (Kaiser Wilhelm Society for the Promotion of Research,
hereafter KW Society), founded in 1911, was to supplement the research
capabilities of the universities. It differed, then, from the Imperial Physics
Institute in that its main function was to promote fundamental research.
The various institutes which were founded under the umbrella of the KW
Society differed from the universities since, in the first place, they provided research time and facilities unencumbered by the need to teach.
Secondly, they allowed for a degree of disciplinary specialisation that was
not easily accommodated in the traditional university. The variety of
institutes founded in the years before World War I indicates the disciplinary range the KW Society covered: Electrochemistry (1911),
Chemistry (1911), Biology (1912), Experimental Therapy (1912),
Labour Physiology (1912) and Coal Research (1913).[15] Their titles
suggest, too, that the aspiration to be an institution for pure research
was modified by the need to support German industry (and to attract
private funding). As with the physics institute, the aim was to bring
together government and private resources. Before the war, the government provided little more than the land.[16] Significantly, the international
models for this German initiative included US research institutions
funded by such private philanthropic bodies as the Carnegie Institute or
the Rockefeller Foundation.[17]

Predictably, however, the amount of private funding fell short of
expectations. Another difficulty for the KW Society was its ambivalent

---

[13] Walker, 'Twentieth century German science', p. 796   [14] Cahan, *Institute*, p. 3
[15] Walker, 'Introduction', p. 242   [16] Johnson, 'Academic chemistry', p. 522
[17] Beyerchen, 'On the stimulation', p. 157

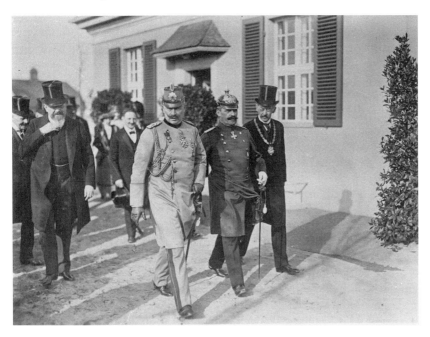

6.1 Opening of the Kaiser-Wilhelm-Institut in Berlin-Dahlem, 1913,
From right: Adolf von Harnack, Friedrich von Ilberg, Kaiser Wilhelm II,
Carl Neuberg, August von Trott zu Solz.

relations with the universities. The KW Society may have been set up
to supplement the research output of the universities, but it lacked
their accrediting privileges. The senior staff of the KW Society were
recruited from the universities, but did not have the rank of professor –
hence, they did not have the right to examine doctorates. For junior
staff whose most likely long-term career prospects lay in the universities, joining the KW Society could be a cul-de-sac, hence recruitment of junior staff could be a problem.[18] Nonetheless, the Society
went some way to addressing some of the universities' deficiencies.
These included their reluctance to embrace applied research, and their
opposition to the establishment of new disciplines in the face of the
universities' attachment to a traditional body of specialisms. For
German science, it meant that there was greater opportunity to establish a research agenda in relatively new areas.[19] The German state
may not have invested much money in enabling this to happen, but it

[18] Johnson, 'Academic chemistry', *Isis*, p. 522   [19] Szöllösi-Janze, 'Science', pp. 342–3

did invest imperial prestige in the KW Society, thus helping to protect its position from more traditional bodies such as the universities. The universities, however, with their combination of teaching and research, remained the chief institutions for the creation and dissemination of new knowledge, and care was taken to avoid the appearance of competition from the Kaiser Wilhelm Society or the Imperial Physical-Technical Institute.[20]

The pattern of scientific research in Germany in the decades up to World War I reflected the changing political structures of the period. Up to unification in 1871, research was largely the province of the universities with support from the different states. The ethic of research had been well established within the universities and helped to provide the impetus for a continuing respect for scientific innovation. After unification, the federal government became increasing involved and reflected the change through the foundation of new institutions, notably the Imperial Physics Institute and the Kaiser Wilhelm Society and its various institutes. By the time of unification, too, German industry had reached new heights, largely through the ability to combine science with technological process. Eventually, indeed, German industry was to surpass the levels achieved by Britain and France. Science contributed to industry but so, too, did industry contribute to science through industrial laboratories, or, in some cases, support for the imperial institutes for research. The federal government itself provided only limited funds for research but helped to facilitate the connection between industry and scientific research whether in the universities, the technical institutes or the imperial institutes. Though the embodiment of a conservative military elite, the royal family played a significant role in this drawing together of the major strands of German industrial might. By such steps was the German state transformed into a major power, in large part because of its ability to harness science to the dynamism of the second industrial revolution.[21]

One stimulus for the growth of the German state's relations with science was, then, to foster a fruitful partnership between science and industry in the age of the Second Industrial Revolution. What made a more tangible difference to the lives of the population at large was the way in which the state expanded its role in providing social services. This linked with science at a number of points: firstly, the promotion of medical research, but also the development of forms of political arithmetic that resembled the methods of science in its reliance on order, system and evidence. By the late nineteenth century, the growing body of statistics and mathematical techniques to deal with them was

[20] Joseph Ben-David, *Centers*, p. 103   [21] Gilpin, *France*, p. 22

a stimulus to the development of new forms of medical research, out of which emerged epidemiology.[22]

The emergence of a Sozialstaat (welfare state) had deep roots in Germany's past, and particularly the tradition of cameralism with its statistically-minded oversight of the population and economy. In 1805, for example, the Prussian state established a permanent census bureau that grew in sophistication and complexity, providing figures not only on population levels, but also commercial statistics and annual yield of harvests, together with the medical and educational facilities of the state.[23] This level of expertise was to be passed on to the unified Germany, which was dominated by Prussia. Something of the same paternalism was evident, too, in the way in which German states had taken a much more active role in monitoring and certifying doctors than had other comparable states.[24] It was on such foundations that the Bismarckian welfare state was erected. The motives of the deeply conservative and militaristic Bismarck for doing so were probably various, but one that was politically particularly important was to take the wind out of the sails of his socialist opponents by providing many of the social benefits for which they agitated. Very likely, too, he hoped to bind together the newly unified German population through a uniform system of benefits. The major planks on which this welfare system was built were the Health Insurance Bill of 1883, Accident Insurance Bill of 1884 and the Old Age and Disability Insurance Act of 1889.

Such measures meant that the state had a strong vested interest in the wellbeing of the population, and particularly its medical wellbeing. The rationale for such state involvement was given further momentum from the development of the germ theory of disease thanks to the work of Louis Pasteur in France and Robert Koch in Germany. The German state gave particular pre-eminence to Robert Koch as an outstanding example of the excellence of German science. Koch was already prominent in governmental circles with his appointment as one of the directors of the Medical Office in 1881. Following his isolation of the tubercle bacillus in 1882 and the microorganism which caused tuberculosis, Koch was feted by government, and in 1891 an institute for the study of infectious diseases was created for him at the University of Berlin. This innovation caused friction within the university, with the result that the institute became one of a number of independent government medical research institutes.[25]

[22] Szöllösi-Janze, 'Science', p. 345   [23] von Oertzed, 'Machineries', p. 143
[24] Tuchman, *Science*, p. 121   [25] Beyerchen, 'On the stimulation', p. 154

Bacteriology with its search for ways to prevent infection tended, as Dorothy Porter argues,[26] to favour a centralised administration. In some ways this suited the governmental structure of Germany, though the continued existence of a federal system complicated compliance with general public health measures because of the variations by individual states. Prussia, however, responded to the scientific developments with the creation of full-time medical officers of health in 1899. The need for the newly unified Germany to deal with the problems caused by the rapid growth of cities and rapid industrialisation provided another link between science and the state. Science moulded the state's response with the development of fields such as bacteriology and epidemiology; reciprocally, science was shaped by the actions of the state in providing statistics with medical or scientific import and in state patronage of some scientific fields rather than others. Both sides, however, benefited from the partnership, providing an influential model on which other nations were to draw.

Britain

While Germany was well aware of the need to modernise its economy and to use science to accelerate such change, Britain, as the first industrial nation, was inclined to complacency. It took the shock of German economic progress to help prompt Britain to find ways to combine its traditions as a liberal state with greater support for science. Not only were there economic considerations at stake, for Britain was expanding its machinery of government to deal with the rising levels of population and social need. As political reform gathered pace in the late nineteenth century, there was a growing need to respond to popular demand for social services provided by the state. The scale of these necessarily demanded those with medical or statistical training drawing the early welfare state within the orbit of those with scientific expertise. As the functions of the state expanded, there were moves to redefine the form of liberalism that had been the defining ideology of the British state. The 'new liberalism' of the late nineteenth and early twentieth century envisaged liberalism not in negative terms as the minimising of the powers of the state but, rather, in more positive terms as the state providing the appropriate facilities for civilised existence, including provision of education, medical services and measures for unemployment. The changing character of the British state as its role expanded brought with it changing relations with the conduct of science.

---

[26] Porter, *Health, civilization, and the state*, p. 108

As it had done since its foundation in 1660, the Royal Society remained the most immediate point of contact between the world of science and that of government – though this role was to be modified by the growth of other institutions over the course of the second half of the nineteenth century and early twentieth century. As we saw in the previous chapter, something of a watershed in the relations between the Royal Society and the state occurred in 1849 with the foundation of a government grant to be administered by the Royal Society. It was the first acknowledgement by the state that fundamental research required some support even when it was not directly linked with the machinery of government. After an uncertain start, the grant was made permanent in 1855, though the mechanics of administration fluctuated with the changing relations with Treasury and other key government departments. Between 1850 and 1914, almost 1,000 scientists were supported by this fund.[27] Over time, as Roy MacLeod has shown, the functions of the Royal Society grant were subsumed by other institutions. Scientific research became less exclusively identified with the Royal Society and became increasingly the province of universities, and the role of the Royal Society as an administrator was replaced by the Privy Council.[28] It was a cameo example of the way in which the state's role as a patron of science could begin with the traditional informal methods, but was to expand and become more encompassing as both the needs of science and of the state increased in size and expense.

The Royal Society itself was somewhat ambivalent about the extent to which it should serve the needs of government. In 1868, the Meteorological Committee of the Royal Society took over the functions of the Meteorological Department of the Board of Trade. The change was not a happy one, and eventually the Royal Society members resigned en masse, protesting that the task of weather prediction belonged to government. Informal methods of gaining scientific advice were ineluctably being replaced by more formal government departments. In the long term, a compromise was achieved when in 1919 the Meteorological Office was placed under the Air Ministry, and the Royal Society provided an advisory body to the new Met Office.[29] As this example indicates, the Royal Society remained open to giving advice, though it was of the view that the day-to-day administration of a particular area where government and science converged was the province of the state.

As it had done for much of its history, the Royal Society continued to play a role in scientific exploration. The Royal Society and the British Association for the Advancement of Science joined forces in lobbying for

[27] Alter, *Reluctant patron*, p. 20   [28] MacLeod, 'The Royal Society', p. 323
[29] Hall, *All scientists now*, p. 159

an expedition to explore a hitherto barely known realm of the globe: the ocean bed and the layers above it, and, most particularly, the forms of life this realm supported. The outcome was one of the most successful of all scientific expeditions, the *Challenger* expedition, which set off with six scientists on board in 1872 on a three-and-a-half-year voyage. Appropriately, its commander, Captain Sir George Nares, was both a naval officer and a fellow of the Royal Society. As so often, curiosity-driven scientific research was also to yield practical benefits, since the surveying of the ocean floor that the expedition set in train was useful for the installation of submarine telegraph cables, which transformed communications.[30] In the tradition of its involvement in Ross's exploration of the Antarctic, the Royal Society in 1898 discussed the scientific aspects of a new voyage that helped to provide the scientific foundations for such future Antarctic voyages as those of Robert Scott (1901–4, 1910–13).[31]

The Royal Society, then, might provide advice, but it could not provide the funds or the level of administration that an expansion of government interaction with science involved. Spurred on by the example of Germany and by the growing importance of more scientifically sophisticated chemical, electrical and metallurgical industries, there was growing agitation for government to expand its funding of scientific research. This ran counter to the traditional liberal view that government expenditure should be kept to a minimum, a view held even by some scientists. The prominent astronomer, George Airy, invoked the traditional orthodoxy when he asserted at the British Association for the Advancement of Science meeting in 1851 that 'in Sciences, as well as in almost everything else, our national genius inclines us to prefer voluntary associations of private persons to organizations of any kind dependent on the State'.[32] One attempt to counter this voluntarist view of science was the Devonshire Commission on Scientific Instruction and the Advancement of Science (chaired by the duke of Devonshire) of 1870–5. It saw an urgent need for Britain to modernise and expand the teaching and promotion of science, and recommended a ministry of science and a science advisory council. But these recommendations were in vain, amounting, as they did, to a radical departure from the view that, as far as possible, science should be self-supporting. There were attempts, however, to keep alive at least some of the aspirations of the Devonshire Committee through the Endowment of Research Movement,

---

[30] Reidy, Kroll, Conway, *Exploration and science*, pp. 170–1; MacLeod, 'Discovery and exploration', p. 53
[31] Hall, *All scientists now*, p. 209   [32] Turner, 'Public science', p. 591

which gathered momentum after 1876.[33] Such attempts were unsuccessful in the short term, though as time went on such agitation did lead to the growing acceptance of the need for establishing major scientific institutions, whether with the help of the state or from private sources.[34]

Though the Devonshire Committee's recommendations were not implemented, its very establishment was a symptom of the growing agitation for a more fruitful union between science and the state. Some scientists even claimed that the state had lessons to learn from the methods employed by scientists. If they were employed, they claimed, self-evident benefits for the public good, like the greater support of science, would be more self-evident.[35] The example of Germany came to be increasingly appealed to, as were the needs of the scientifically increasingly sophisticated industries of the Second Industrial Revolution. From the 1870s, there was growing pressure for more scientific expenditure. Science was increasingly portrayed as a means of promoting economic growth and military readiness, major considerations for a state in increasing competition with rival states.[36] Such expenditure meant gaining the confidence of Treasury, the abiding concern of which was to ensure accountability of public funds[37] – not an easy task when funding the uncertainties of research. Eventually, however, Treasury devised forms of accountability that relied on expert scientific advice, thus providing a clearer administrative path for government support for the promotion of science.

Throughout such debates, scientists in the employ of the state continued to perform such routine functions as the assaying of metals at the Mint, the study of explosives at Woolwich Arsenal or the oversight of that great centre of imperial botany, Kew Gardens.[38] Science and the state could converge for specific utilitarian functions, but those agitating for change sought government support for the kind of research necessary to benefit new industries and for Britain to remain a major centre of science. This required an institutional infrastructure to train researchers and to provide the laboratories and other facilities to conduct such research. During the closing decades of the nineteenth century and the period before World War I, such institutions began belatedly to appear. An obvious site for such facilities was the universities, but before 1880 England had only four: Oxford, Cambridge, London and Durham (the same number as Scotland). In 1880, however, the federal Victoria University was established with its headquarters in Manchester, but

[33] Alter, *Reluctant patron*, p. 79   [34] MacLeod, 'Resources of science', p. 160
[35] Turner, 'Public science', p. 594   [36] Turner, 'Public science', p. 593
[37] MacLeod, 'Science and the Treasury', pp. 158–9   [38] Russell, *Science*, p. 246

eventually branch campuses at Liverpool and Leeds were founded. Another federal university, the University of Wales, followed in 1893. The number of universities continued to expand in the early twentieth century: Birmingham (1900), Sheffield (1905), Bristol (1909) – such rapid growth was an index of the growing reach of the state and its willingness to spend public resources, not only on education, but also on the research that, as in Germany, was becoming a related primary function of the university.

A number of more specialist research institutes were also set up with the sanction of the state. In 1899, the Colonial Office established the London School of Tropical Medicine. It was a foundation which reflected the expansive ideology of the late nineteenth-century 'new imperialism', an enthusiastic embrace of empire which was the outcome of growing imperial competition between the Great Powers. The strength of the German example was most evident in the foundation of the National Physical Laboratory in 1900. Its goals were similar to the Imperial Physical and Technical Institute: to establish national standards, especially in such new areas as electricity, and to conduct research in the physical sciences. The example of the German Institute also was evident in the way in which it was envisaged (especially by Treasury) that the National Physical Laboratory would be financed from private sources, though operating under the aegis of government. Indeed, the need to generate income by taking in commissions led to some tension with the industries that had previously handled such work. The hand of Treasury was, however, at times heavy: in 1912, it helped to frustrate such projects as optical research, leaving Britain dependent on German (60 per cent) and, to a lesser extent, French imports (30 per cent) apart from a mere 10 per cent from British sources[39] – a major liability once war broke out. Despite such limitations, the laboratory was seen as a flagship institution signalling a new relation between science and the state: rather hyperbolically, the Prince of Wales referred to it at its opening as 'the first instance of the State taking part in scientific research'.[40] What was (relatively) new was its primary focus on research, with an ethos that merged the pure research of the universities with the applied research of industry.[41] As a government facility, it took over an earlier testing centre at the old Kew Observatory, thus divesting the Royal Society (which had run and financed this facility since 1872) of a task that more naturally belonged to government. The Royal Society, however, continued to

---

[39] MacLeod and MacLeod, 'War', p. 169; Alter, *Reluctant patron*, p. 112
[40] Moseley, 'The origins', p. 236   [41] Moseley, 'The origins', p. 248

oversee the Laboratory, though also with representation of the British Association for the Advancement of Science on the committee.[42]

The German example was also apparent in the foundation of the Imperial College of Science and Technology as a constituent part of the University of London in 1907.[43] The Germans had shown the value of technological institutes of higher education, and the Swiss had also provided a successful model with the Swiss Federal Institute of Technology in Zürich founded in 1854. A distinctive British note was added with the adjective 'imperial', indicating the extent to which the college was meant to serve the British Empire as a whole. This was a reflection of the more assertive imperial sentiment that came to the fore with the increasing competition between the Great Powers. Tellingly, the College's motto was 'Scientia Imperii Decus et Tutamen' ('Science is the pride and shield of empire').[44] The Imperial College drew together some of the existing institutions for advanced education in science, chiefly the Royal College of Mines and the Royal College of Science (which had grown out of the Royal College of Chemistry). The effect of these amalgamations, then, was to bring greater focus and prestige to the study of science and technology by placing existing instruction in a university context. Such a move attracted support from industry,[45] and hence improved facilities. It also underlined the extent to which the College was expected to support itself from private sources. The College's foundation was a clear statement, then, of the importance of science in both the national and imperial arenas.

New institutions dedicated to the promotion of science, like the National Physical Laboratory or Imperial College, were a visible indication of the growing attention that science was receiving from the state. Less immediately visible, but important nonetheless, was the way in which scientific standards of evidence, using quantitative measures where possible, were increasingly being applied to problems of government. The most outstanding example of this was in the area of public health, where the development of statistical studies based on the census and other surveys provided the data for drawing conclusions about patterns of disease. Such a process was greatly assisted by the establishment of the Registrar General's Office in 1836 and the accompanying compulsory registration of births, marriages and deaths. State-provided statistics, then, allowed the growth of sciences like epidemiology. Indeed, in 1850 the Epidemiological Society of London was founded to attempt to study

[42] Alter, *Reluctant patron*, p. 137   [43] Alter, *Reluctant patron*, p. 158
[44] Kumar, *Science and empire*, p. vii   [45] Alter, *Reluctant patron*, p. 46

the dynamics of epidemics.[46] Faced with the congestion and high death rates of the cities of the Industrial Revolution, increasing attention was paid to the way in which diseases were transmitted. The availability of statistical evidence enabled John Snow in his 1849 study, *On the mode of communication of cholera*, to show that cholera was a waterborne disease. However, it took successive outbreaks of disease and a growing number of studies of their causes before authorities could be persuaded to devote the resources to implementing such basic measures as ensuring a clean water supply.

Dealing with major social problems, from the regulation of factories to the development of building standards, required from the Victorian state both time and successive legislation. Those involved in implementing existing regulations urged the need for further such regulations to deal with the scale and complexity of the problem. Such was the case, too, with public health, so that the 1848 Public Health Act establishing a General Board of Health was but the first in a series of acts intended to grapple with the problem. A fundamental problem with all such legislation was that it involved an expansion in the powers of the state that ran contrary to the prevailing liberal view that such powers should be circumscribed as much as possible. One of the countervailing forces that enabled the state's power to increase was the growing respect for forms of scientific evidence, most notably in the form of statistics. Another was the growth of new formulations of liberalism, such as the new liberalism, which, in contrast to the traditional negative depiction of the state's power over the individual, emphasised the need for the state to provide the individual with fundamental services such as health and education. The new liberalism, which was influential in the decades before World War I, also overlapped with social imperialism, which combined imperial expansion and consolidation abroad with social reform at home.

Gradually, then, the British state expanded its powers so that, as Dorothy Porter points out, the liberal constitution which framed political life increasingly existed side by side with an ever more watchful bureaucracy.[47] In the name of science the state compelled its citizens to comply with such rules as the 1851 Act, which made some vaccinations for various diseases compulsory. This was later supplemented by the 1871 requirement that vaccinations be registered by a vaccination officer, and the 1889 Infectious Disease Act that stipulated compulsory notification of all such diseases. From public health measures concerning the population at large, the British state turned to medical insurance for the

---

[46] Porter, *Health, civilization, and the state*, p. 69
[47] Porter, *Health, civilization, and the state*, p. 111

individual with the National Insurance Act of 1911 – a pioneering measure in laying down the outlines of a framework of social services that owed much to the example of Bismarckian Germany.

Under the provisions of this Act, a small proportion of the medical insurance money collected was dedicated to medical research. This necessitated a state body to coordinate such research, which took the form of the Medical Research Committee, established in 1913. This proved to be a significant milestone in the relations between the state and the scientific community in Britain. The traditional fear among scientists that state funding would mean state control was alleviated by the way in which the Medical Research Committee acted as a buffer between the scientists and the state. Scientists were not bound by directives from above but could pursue their research as seemed appropriate (and manageable given available resources). It was an innovative approach that maintained some of the British liberal traditions and served as a model for later British bodies, as well as those abroad.[48]

Closely allied with the state's role in dealing with public health was legislation dealing with what would be called environmental issues. A pioneering measure in this area was the Alkali Act of 1863, originally implemented to prevent discharge of gaseous hydrochloric acid into the atmosphere. To reinforce the point about the way in which Victorian social legislation often evolved as the scale of the problem became more evident, the original 1863 Act was followed by Acts of 1874, 1892 and 1906. Behind such successive legislation lay the influence of the professional inspectors who, as Roy MacLeod argues, were leading the British state into an expansion of its responsibilities by using the scientific expertise of civil scientists to protect both the general population and industry.[49]

When World War I struck, then, Britain had done much to integrate science into the workings of the state in a way which served it well during the war. Government had recognised that it had much to gain from an alliance with science and that science could not be supported by private sources alone. Such a message was driven home by foreign competition, as Britain fell behind in modern science-related industries. From 1880 to 1913, Britain's proportion of the global export market for manufactured goods shrunk from 41.1 per cent to just under 30 per cent while Germany's increased from 19.3 per cent to 26.5 percent, with the figures for the USA going from 3 per cent to 12.6 per cent.[50] To rectify this decline, Britain had instituted new forms of government involvement

---

[48] Alter, *Reluctant patron*, p. 172   [49] MacLeod, 'The Alkali Acts', p. 86
[50] Alter, *Reluctant patron*, p. 106

with science, from public health and alkali inspectors to support for the National Physical Laboratory. The need for greater scientific education was addressed with the foundation of the Imperial College of Science and Technology and the foundation of new universities. The degree of symbiosis between university research and the needs of industry remained a concern, however, prompting the formation of a department of the Board of Education to improve the level of research at British universities along with its relevance to industry.[51] Importantly, Britain had devised models for involving scientists in state-directed research in a way that allayed their fears that they were losing their professional independence. Most notable in this regard was the role of the Medical Research Committee as an in-between body – though the way in which the National Physical Laboratory straddled the worlds of industry and university research was also important. Such innovations were to be built on further during World War I, providing the framework for British science policy that was to remain in essentials until after World War II.[52]

British Imperialism

The British state was not only expanding in its power and possibilities at home, but was also doing so abroad. The late nineteenth and early twentieth century was the heyday of the European empires, as the Great Powers drew on their industrial might to challenge each other's hold on different sectors of the globe. Just as science played a role in consolidating the reach and power of the state at home, so too was it drawn on as one of the supports of empire. As Sujit Sivasundaram points out, empires and science had much in common: both gave epistemological weight to ordering knowledge, system and classification.[53] Science used the networks of empire both to collect specimens and to distribute them. In doing so, it used institutions such as botanical gardens or observatories the location of which was often determined by the vagaries of colonial rule. Above all, science carried authority both at home and abroad and provided a justification for imperial rule, as science was viewed as a means of economic and intellectual improvement. Science could also work as a means of delegitimating local practice in fields as diverse as agriculture or medicine. Indigenous experience was disregarded if it did not comply with the canons of scientific rigour that were prescribed by the imperial state.[54] Such authority was reinforced by such advances in technology as the network of telegraphic cables which tied empires together on a global

[51] Alter, *Reluctant patron*, p. 204   [52] Alter, *Reluctant patron*, p. 6
[53] Sivasundaram, 'Sciences', p. 154   [54] Bonneuil, 'Development'

scale, or, in places like India, the expansion of railway systems[55] that further consolidated the growth of a distinct state.

Within Britain the growing intermeshing between the metropolitan and imperial state in the late nineteenth and early twentieth century was reflected in the further evolution of Kew Gardens. This, as discussed in Chapter 5, was made a formal department of state in 1840, having been previously a royal possession. Kew acted as the great central imperial clearing house of botanical specimens and the place to nurse botanical specimens until they were capable of being replanted in an appropriate colonial locale. A notable example was the way in which, at the initiative of the Viceroy of India, rubber plants were transferred from Brazil to Kew Gardens, and thence to Singapore and Malaya.[56] Kew's power was an artefact of empire, especially as the network of local botanical gardens on which its potency rested was largely paid for by local colonial governments.[57] As the British state became more preoccupied with the uses of science, Kew expanded its scientific reach. In 1876, it was enhanced by the addition of the Jodrell Laboratory for the study of plant physiology and cytology.[58] In 1902, the close association between Kew and the imperial state was made manifest by the appointment of its director as botanical advisor to the Colonial Office.[59]

This was indicative of the increasing interest the Colonial Office was taking in science as a means of bringing greater prosperity to the empire (or, at least, those who ruled it). This was particularly marked during the period when Joseph Chamberlain served as Colonial Secretary between 1895 and 1903, creating an alliance between science and development that remained influential until the end of the British Empire.[60] It was Chamberlain who appointed a Royal Commission, an accumulation of ordered and authenticated information that resembled a scientific project, to investigate the decline in the sugar production of the West Indies. The outcome was an expansion of the scientific establishment with the foundation in 1899 of an Imperial Department of Agriculture with its headquarters at Barbados.[61] With this precedent established, such departments of agriculture spread throughout the empire. The network of botanical gardens over which Kew presided was to act as a bridgehead for the further expansion of science in an imperial context. Of the twenty-six colonial botanical gardens linked to Kew in 1900, all were to serve as

[55] Headrick, *Tools*, pp. 163, 181   [56] Headrick, *The tentacles*, p. 252
[57] Storey, *Science and power*, p. 63   [58] Brockway, *Science and colonial expansion*, p. 86
[59] Headrick, *The tentacles*, p. 215
[60] Hodge, 'Science and empire', pp. 13–14; Worboys, 'Science and the colonial empire', pp. 13–27
[61] Headrick, *The tentacles*, p. 216; Storey 'Plants, power and development', pp. 109–30.

PALM-HOUSE, KEW GARDENS.

6.2 Palm House at Kew Gardens, from 'Collins' Illustrated Guide to London and Neighbourhood' (1871). The iconic Palm House was designed by Decimus Burton, built by Richard Turner and dates from 1844 to 1848.

the nucleus for a department of agriculture.[62] While Kew promoted such a development, the consequent decentralisation was to weaken its imperial power.[63]

Another indication of the Colonial Office's greater interest in linking together British scientific resources with imperial rule was, as we have seen, the foundation of the London School of Tropical Medicine in 1899. In the same year the Liverpool School of Tropical Medicine was also founded. Such developments in imperial medicine could also be observed in other empires, with the relatively small and belatedly founded German empire acquiring a similar institution with the foundation of the Hamburg Institute for Naval and Tropical Diseases in 1901. The London School, in turn, established the Tropical Diseases Research Fund, which was linked both with the British scientific establishment, with the support of

[62] Worboys, 'Science and the colonial empire', p. 18
[63] Storey, *Science and power*, p. 97

the Royal Society, and the empire, through support from India and the Crown colonies. The demands of empire, however, could be a source of conflict with local bodies such as the British medical profession, which resented the degree of control the London School of Tropical Medicine exerted over many of their members. The London School's patron, the Colonial Office, however, had considerable leverage in matters medical by having control over some three hundred positions.[64] Tropical medicine, to a degree, made European empires possible by helping to maintain colonial officers alive and in reasonable health. It was not until malaria was largely conquered with the use of quinine, for example, that European rulers took possession of much of Africa. Tropical medicine also gave colonial rulers authority as indigenous peoples saw the power which scientifically based medicine enabled them to exert over disease. Ronald Ross's discovery of the way in which malaria is transmitted via mosquito bites in 1898 was followed in 1899 by his transfer from India back to England and a post at the Liverpool School of Tropical Medicine. There, his achievement became a matter of national pride as a notable instance of the way in which science could be advantageously linked with the consolidation of empire.[65]

The powers that an imperial state exerted made compliance with government demands for information often more readily obtainable than at home in the metropolitan power. Such data was, as we have seen, one of the state's major sources of power, making it possible to plan and execute policy in a scientifically orderly manner. Imperialism stimulated such data collection both at home and abroad.[66] In India, the population was classified in a more comprehensive way than in Britain, with detail, for example, on religion – or what the British understood by religion. Mapping proceeded early in the Raj, reaching a considerable level of scientific sophistication with the establishment in 1818 of the Great Trigonometrical Survey, which oversaw survey operations in general, including those related to revenue matters.[67] The demands of public health led to imperial states trying to change the social habits of those it ruled both at home and abroad. In an imperial setting, particular attention was devoted to centres of control where there was a European population such as settlements, plantations and mines – especially as these were likely to be economically productive areas. This strengthened the association between medicine and imperial rule,[68] which was also underlined by the neglect of Indian medical traditions or local drugs in the

---

[64] Petitjean, 'Science and the "Civilizing Mission"', pp. 126, 128, 151, 154, 157
[65] Petitjean, 'Science and the "Civilizing Mission"', p. 124
[66] Cohn, *Colonialism*, pp. 3–4   [67] Kumar, *Science and the Raj*, p. 74
[68] Worboys, 'Public and environmental health', p. 155

teaching of medicine even within India. Medicine, however, received less attention than sciences such as botany or geology with a more obvious economic return. The experience of dealing with epidemics in Britain and the growing intrusion of the imperial state in the late nineteenth and early twentieth century led to more systematic enquiries that strengthened the hand of government. In 1861, a major cholera epidemic led to the estab-lishment of a cholera commission, the first such scientifically based mass study. Others followed, such as the Plague Commission Report of 1904.[69]

The association between science and empire could transcend the for-mal bounds of empire, particularly in the late nineteenth century as British influence around the globe became more pronounced. British science made inroads into China, which fell within Britain's informal empire (with the minor exception of the small Crown Colony of Hong Kong). Apart from a botanical garden in Hong Kong and the British Consular Service, natural history collectors were reliant on indir-ect agents of empire, such as missionaries or paid indigenous collectors. Success largely depended on securing the cooperation of such indigenous collectors, who possessed a degree of local knowledge beyond a Westerner. The process of drawing such natural history information into European systems of knowledge meant, however, a distancing of such knowledge from its indigenous context and an epistemological expansion of Western understandings of nature at the expense of locally based ones.[70]

Expansion of the British state in the late nineteenth and early twentieth century had consequences both at home and abroad. Indeed, the two axes of British power were connected with greater imperial power increasing the British state's resources and authority at home; conversely, its greater facility as a 'data state' at home widened its bureaucratic reach in ruling abroad. Reliable data derived from surveys, censuses and official enqui-ries became the coinage of rule and buoyed the confidence of rulers that their policies had a rational and, indeed, scientific basis. The increasing intertwining of the domestic and the imperial also meant the growth of key scientific institutions such as Kew Gardens, until local scientific establishments came to weaken its centralising control. The Royal Society from its foundation in 1660 saw itself as drawing on the Empire as a source of scientific information.[71] This tradition it carried further in the late nineteenth and early twentieth century with the Royal Society's

---

[69] Kumar, *Science and the Raj*, pp. 166, 167, 178
[70] Fan, *British naturalists*, pp. 61, 83, 89   [71] Gascoigne, 'The Royal Society'

increasing representation in imperial affairs.[72] The Royal Society, for example, established an Indian Advisory Council to assist the government of India.[73] British medicine acquired a more imperial overview with the establishment of institutes specifically devoted to tropical medicine. Such developments were a reflection of an expanding state bringing to bear on the functions of government both at home and abroad the possibilities of science.

France

Distracted and divided by the conflicts over the character of the post-Revolutionary state which had persisted since the Revolution, France was slower than Germany and Britain to expand the boundaries of the state. In any case, the governmental structures laid down in the Napoleonic regime largely persisted, inhibiting innovation. Reforms in public health, for example, were slower to emerge than in Britain, though they gained momentum after the institution of the Third Republic in 1871 with its pledge to improve the health of the population. This ideological crusade was given scientific force through the development of bacteriology, and especially the discoveries of Louis Pasteur. Nonetheless, it took some time to overcome the entrenched conservatism of much of France: a comprehensive law requiring all areas to enforce vaccinations and their notification and to deal with unsanitary sites was introduced in 1893, but was not finally enacted until 1904.[74] Consistent with this, too, France, was slower to implement measures comparable to the welfare reforms of Germany and Britain.

One area of expansion of the state's role was in the provision of scientific education. To bring together science and industry, the state promoted a wide range of technical institutes[75] that emulated those of Germany. University science faculties were also strengthened and encouraged to work with local industry.[76] Such alliances meant that the pace of French scientific research was to a large extent set by the level of French industrial innovation. German industry, however, considerably outpaced that of France, especially in such key areas of the Second Industrial Revolution as the chemical and metallurgical industries. The result, then, was that France continued to lag behind Germany in scientific and industrial research.[77]

[72] Roy Macleod, 'Imperial reflections', p. 167   [73] MacLeod, 'Scientific advice', p. 345
[74] Porter, *Health, civilization, and the state*, pp. 101–3
[75] Day, 'Education for the industrial world', p. 144
[76] Shinn, 'The industry, research, and education', p. 140
[77] Zwerling, 'The emergence of the École normale superieure', p. 60

The pace of scientific innovation was largely set by competition with other nations, and especially Germany. Even before the Franco-Prussian War of 1871 there was an awareness of the need for France to respond to the German challenge.[78] In 1868, there was an initiative to improve the training available in universities through the creation of the École pratique des hautes etudes (Applied School of Higher Studies).[79] What was innovative about this institution was its focus on training in research techniques with laboratory work in the sciences and seminar participation in the humanities.[80] Such training was offered in a range of disciplines: mathematics, physics and chemistry, natural sciences and physiology, and historical and philological sciences (with religious sciences being added in 1886). Along with postgraduate education, its main mission was to provide funds to appropriate university centres, thus disseminating its methods.[81] Of the existing Grandes Écoles, the one which responded most adroitly to the changing needs of the French economy and its industrialisation was the École normal superieure (Superior Normal School), which came to eclipse the École Polytechnique. Founded as a training institution for senior high school teachers, the Normal School came to provide a rigorous training in the physical sciences that provided the basis for an alliance with industry.[82]

Defeat in the Franco-Prussian War brought much greater impetus for the French state to keep pace with Germany by expansion of its scientific establishment. Critical, too, was the nature of the state that followed the defeat in the Franco-Prussian war of 1870–1. The leaders of the new Third Republic had a fervent allegiance to the faith of the Revolution and, with it, the values of reason and science and, negatively, its strident anticlericalism. Probably in no other European country was science so closely intertwined with the ideology of the state.[83] One outstanding embodiment of Revolutionary scientific rationality was the French metric system. It was appropriate, then, that, following the Metre Convention of 1875, the Bureau international des poids et mesures (International Bureau of Weights and Measures) was established in 1876 at Sèvres, where, to this day, it enjoys an extraterritorial status. The foundation of the Institut Pasteur in 1887 showed an openness to new forms of institutional structure as well as a commitment to furthering research. There Pasteur and his successors could surround themselves with fellow specialists in the biological sciences working on major collaborative projects.

[78] Fox, 'Scientific enterprise', pp. 464–5   [79] Paul, *The rise of scientific empire*, p. 5
[80] Zeldin, 'Higher education in France', p. 77
[81] Guerlac, 'Science and French national strength', p. 87
[82] Zwerling, 'The emergence of the École normale superieure', p. 33
[83] Paul, *Rise of scientific empire*, p. 38; Fox, *Savant and the state*, p. 1

Unusually in France, the funding was largely private, though with some funding from the state. The concept of a privately-financed scientific research institute was to have considerable influence, with the Rockefeller Foundation later drawing on the French model.[84]

Though the Institut Pasteur was a success, it was evident that if France was to maintain an international reputation in scientific research, private resources would not suffice. What was needed was a governmental scheme to supplement the resources available to researchers. This was particularly true in the science faculties of the centralised national university, in which there were very limited resources available for research. The beginnings of such a scheme came with the foundation in 1901 of the Caisse des recherches scientifiques (Scientific Research Fund), which supported scientific research in the faculties with strong priority for areas with economic potential.[85] Such measures bore fruit in a very marked rise in the research productivity of the universities,[86] institutions which had been left in a fragmented and demoralised state by the Napoleonic reforms that created one single university for the whole of France. The faculties were, then, the beneficiaries of state largesse at a time when the traditional central point of French scientific research, the Academy of Sciences, was being relatively deprived of funds. Under the Third Republic, particularly, the Academy was looked upon as a bastion of conservatism in contrast to the faculties, which were more likely to be dominated by those with a robustly republican outlook. One small concession which the Academy finally made in 1913 to offset the centralisation in science, to which it had greatly contributed, was to establish a section for scientists outside Paris.[87] Perhaps the provinces might be receiving more attention, but the system for distributing research funds was in keeping with the bureaucratic traditions of the French state that had been the foundation, too, for the considerable expansion in scientific education. The expansion of both the teaching and research of science, then, was closely allied with the central state, taking even further the close intertwining of science with the French state.[88]

French Imperialism

The second half of the nineteenth century was a time of expansion for the French empire. Its first empire had been largely lost in the wars with the

---

[84] Guerlac, 'Science and French national strength', p. 88   [85] Gilpin, *France*, p. 154
[86] Paul, *Rise of scientific empire*, p. 3
[87] Crosland, 'The French Academy of Sciences', pp. 84, 102
[88] Fox, *Savant and the state*, p. 2

British up to 1815, while its second principally dated from the 1860s (with the exception of Algeria) and included the colonisation of Senegal, Indo-China and parts of the Pacific, along with the consolidation of rule in North Africa. Given the prominence of science in French culture, it was naturally reflected in the character of French colonial rule.[89] Indeed, Daniel Headrick writes of France (and Germany) in this period that 'the institutionalisation of the sciences was part of the larger process of state formation and its expansion . . . The sciences could thus become an "official" agent in the debate about the justification of imperial engagement.'[90] In France, the spread of science and technology was seen as a central element of the 'civilizing mission' that was at the core of the justification for the spread of imperialism. The French state helped to strengthen the association between science and imperialism by its sponsoring of scientific discovery missions in conjunction with the military.[91]

Though French imperialism and science were closely connected, the French National Museum of Natural History never served as a centre of imperial botany in the way that Kew Gardens did (or the Dutch gardens in Java at Buitenzorg (founded 1817)). The French were less committed to economic returns than the British or the Dutch, though the Museum reaped considerable scientific profit from the spread of empire. Late nineteenth-century imperial expansion brought a greatly increased scientific productivity to the Museum, even though it had fallen behind other institutions (such as the Paris science faculties) in the competition for state funding. The wealth of samples from colonial sources led, however, to an emphasis on unfashionable descriptive natural history, which placed its scientific output on the periphery of the scientific interests of the day.[92] Foreign competition did lead to a greater interest in the economic potential of colonial natural history, with the result that a Jardin d'essai (Experimental Garden) was established in 1899, which the Museum attempted to control. Another indication of greater engagement with colonial natural products was the establishment in 1900 by Paul Doumer, Governor-General of French Indochina from 1897 to 1902, in 1900 of a Mission permanente d'exploration scientifique (Permanent Mission of Scientific Exploration), but this was closed in 1908.[93]

The National Museum of Natural History also exerted some indirect influence through the Société zoologique d'acclimatation (Zoological Acclimatisation Society), an organisation founded out of the Museum

[89] Headrick, *The tentacles of progress*, p. 223   [90] Headrick, *The tentacles of progress*, p. 44
[91] Petitjean, 'Science and the "Civilizing Mission"', p. 119
[92] Limoges, 'The development of the Museum', pp. 236, 240
[93] Headrick, *The tentacles of progress*, p. 229

in 1854. As its name suggests, its aim was to acclimatise animals from the colonies in the metropolis and vice versa. To that end it erected a zoo at the Bois de Boulogne. It was essentially a private body, though it received some state funds. It was part of a growing number of specialised scientific bodies that meant that the state's traditional reliance for advice on the Academy and the Museum was now lessened as the pool of potential advisers grew larger.[94] Science had become so important to the French state both at home and abroad that the number of organisations needed to foster it and advise government had increased substantially. Such growth was an index of the place of science in French national life, particularly under the Third Republic, when science formed a major part of the national ideology.

The United States

In the second half of the nineteenth century the United States was, in the most literal sense, an expanding state. Whole tracts of the Midwest where once the indigenous inhabitants hunted the buffalo now came under the plough. Inevitably, the expansion in the area governed brought with it an increase in the size of government and, with it, greater attention to the possibilities of scientific investigation on the part of the state. The United States was also expanding economically, moving from a largely agricultural society to one heavily engaged in industry, and particularly the industries of the Second Industrial Revolution: chemical, electrical and metallurgical. Along with such trends towards a widening sphere for an alliance between science and the state went a traditional fear of allowing the federal government to accrue too much power. This acted as a brake on most scientific activities that did not have a clear utilitarian rationale. The general political consensus was that the pure abstract sciences belonged in the universities, and particularly the private universities, which were independent of government.[95]

Any action on the part of the federal government in the middle decades of the nineteenth century was constrained by the drift towards the Civil War (1861–5) and the outbreak of the war itself. Given the suspicion on the part of the Southern states of the federal government, legislation affecting the nation as a whole often did not survive its passage through Congress. Bills which were thus delayed were, however, able to be passed once the Southern states had withdrawn from Congress. The most notable of these in relation to science was the Morrill Land Grant Act of 1862, establishing state colleges specialising in practical skills relating to

[94] Osborne, *Nature, the exotic*, pp. xiv, 1, 149   [95] Kevles, *The physicists*, p. 11

agriculture and mechanics. In the same year, an act establishing a federal department of agriculture was also established. This was not altogether simply a coincidence, since many of the personnel for the department of agriculture were drawn from the graduates of the state colleges. The colleges, too, were to provide the base for agricultural research stations that became scientifically more important after the Hatch Act of 1887 widened their scope and, with it, the level of government funding.[96] The federal department of agriculture was to grow rapidly in size and significance, catering as it did to the needs of so much of the population. No other country was to match its size, and it was to become the largest research agency in the country by 1916.[97]

The Civil War did little to strengthen an alliance of science with the state as a means of increasing its range and potency of weapons.[98] It proved to be a war fought largely with traditional armaments. Scientists invoked the traditional rhetoric that science was above war, and established the National Academy of Sciences in 1863 as a scientific reflection of a unified country (though Southern scientists were barred by a much-contested loyalty oath that was dropped in 1872).[99] It sought to distance itself from the aristocratic traditions of European academies by emphasising its credentials as a meritocratic body.[100] From the beginning it emulated the Royal Society of London, however, in having as a primary goal the provision of advice to government. Thus, its charter insisted that it should 'whenever called upon by any department of the Government, investigate, examine, and report upon any subject of science or art'.[101] In practice, the Academy was rarely troubled until World War I, but it did provide some scientific advice. It gave detailed scientific instructions, for example, to the Charles Hall expedition to the North Pole in 1871–2, and oversaw the analysis of the data and collections that the expedition returned. The aim of the expedition was to be the first to reach the North Pole, and this forlorn hope, with its intimations of glory for the United States, prompted a congressional grant of 50,000 dollars, a rare subsidy for scientific enquiry. The Academy also had a role in planning the Geological Survey in 1878 and a forestry system in 1896.[102] The Academy itself did not secure any congressional funding until 1941.[103] It was, however, a voice for the interests of science, a role complemented by an earlier and very different organisation, the American Association for the Advancement of Science,

---

[96] Rosenberg, 'Science, technology, and economic growth', p. 182
[97] Dupree, *Science*, p. 150   [98] Dupree, *Science*, p. 108   [99] Dupree, *Science*, p. 143
[100] Devles, 'On the flaws', p. 149
[101] Cochrane, *The National Academy of Sciences*, p. 134
[102] Cochrane, *The National Academy of Sciences*, pp. 112–3, 127
[103] Bruce, *The launching*, p. 304

founded in 1848. While the Academy elected a restricted number of eminent fellows, after 1866 the American Association was open to all. Like the British Association for the Advancement of Science, a natural model for its American counterpart, it moved from city to city for its annual meetings, seeking to widen the constituency for science among the general public.

In the aftermath of the Civil War, much of the government's interest in science continued to centre on the mapping and exploitation of the resources that westward expansion had brought with it. More and more the role of exploration of new territories passed from the army to organisations with a more formal scientific foundation notably the Geological Survey founded in 1879.[104] The chief responsibility for geological surveying passed from the states to the federal government, state surveys having been set up earlier in the century (with Massachusetts providing the first state-sponsored survey in 1830). The Corps of Topographic Engineers, which had taken such an active part in antebellum exploration of the West, was a casualty of the Civil War, which had much disrupted its activities.[105]

Another sign of the waning role of the armed forces in what were becoming civilian precincts was the transference of the meteorological service from the army to become the Weather Bureau of the Department of Agriculture in 1890. The Coast Survey had long been vulnerable to incorporation into the Navy, which had traditionally supplied some of its personnel. In 1871, it had widened its scientific remit to include geodetic surveys – congressional approval for such a move came in the form of its formal name change to the Coast and Geodetic Survey in 1878.[106] By 1900, the Survey was completely civilian in character and no longer under the shadow of the Navy.[107] Such shifts from military to civilian control implied the normalisation of many of the scientific functions of the state. No longer were such functions considered so removed from other departments of state that they needed to be handed over to the armed forces in the absence of any more obvious bureaucratic home.

The possibility of expanding the role of the armed forces at the expense of the civilian scientific departments had also been considered and rejected by the Congressional Allison Committee of 1884–6 (chaired by Senator W. B. Allison). The Committee was prompted by demands for retrenchment, especially by those who opposed in principle the patronage of science by the state. Such was the view, for example, of Representative

[104] Dupree, *Science and the emergence*, p. 214   [105] Bruce, *The launching*, p. 296
[106] Wraight and Roberts, *The Coast and Geodetic Survey*, p. 26
[107] Manning, *US Coast Survey*, p. 149

Hilary Herbert of Alabama, who asserted that science had been stifled by such patronage as that bestowed by Louis XIV. Such views also found voices within the scientific establishment, notably that of the naturalist Alexander Agassiz, of Harvard.[108] The outcome of this Committee established 'to secure greater efficiency and economy of administration of the public service'[109] indicated the growing reconciliation of the federal government with the increasing role of science in its bureaucratic structures. Its establishment reflected unease at the growth of such scientifically based departments as the Geological Survey and the Coast and Geodetic Survey, but in the end vindicated their existence by not recommending any substantial changes. It also prompted a committee of the Academy to suggest the creation of a separate department of science to oversee the use of science in all the workings of government. This was not endorsed by the Committee[110] however, and the proposal lapsed as being an impractical expansion of bureaucracy. Surviving the Committee more or less intact was an indication of the extent to which scientific agencies were being accepted as an integral part of the American state.

The role of agencies such as the Geological Survey and the Coast and Geodetic Survey, with their surveys of American territory and its coasts, came, then, to be an accepted part of the government's activities. More controversial, however, was the extent to which the state should assist industry, which was thought to be self-sufficient with increasingly important laboratories of its own. Some tasks could, however, only be performed by government. This was notably true of the setting of standards, particularly in the electrical industries, one of the new boom industries where science merged with technology. For a time, such needs were met by the Office of Weights and Measures within the Coast and Geodetic Survey, but it proved too small for the ever-growing task. Consequently, such pressures led to the foundation of a new institution which brought together science and the state: the National Bureau of Standards, under the control of the Treasury, in 1901.[111] The Bureau marked state recognition of the increasing role which science was now playing in industry, as the Western world passed through what Edward Lyton describes as 'the scientific revolution in technology'.[112]

The growing number of bureaucratic agencies dealing with science brought with it an increasing impetus for the state to govern according to the canons of evidence and mathematical reasoning that were central to

---

[108] Miller, *The political economy*, p. 106   [109] Dupree, *Science*, p. 215
[110] Cochrane, *The National Academy*, pp. 146, 148
[111] Cochrane, *The National Academy*, p. 171   [112] Lyton, 'Mirror-image twins', p. 210

science. For all its opposition to big government, the United States was moving to be a 'data state'. Along with the various agencies to support exploration, agriculture and industry, in 1903 the US Census was given a permanent bureaucratic home in the Department of Commerce and Labor, this providing a secure basis for the generation of social statistics of increasing variety and significance. John Wesley Powell, the dynamic head of the Geological Survey, summed up the view of the state held by many of his fellow Washington scientific senior bureau chiefs when he described it as 'the grand unit of social organization' – a view that naturally provided an incentive for amassing what data was available to achieve orderly government.[113]

The United States was, however, slower than Germany or Britain to use government data as an aid for dealing with the problems of public health in an age of rapid industrialisation and urbanisation. Fear of the power of the central state, combined with an emphasis on states' rights and the need for local initiative, inhibited federal government action.[114] As had happened in a number of areas where scientific expertise was needed but government held back from creating a new department, the problem passed to the armed forces. Lack of any other federal agency to deal with major public health issues led to major expansion of the Marine Hospital Service, which, though not formerly a part of the army or navy, was closely affiliated with them and adopted similar mores. This Service had been established in 1798 to tend to sick and injured members of the navy and the merchant navy. Its military character was reaffirmed in the reforms of 1871, when its growing size led to the appointment of its first supervising surgeon, John Maynard Woodworth. He adopted a military model with uniforms for physicians along with a system of examinations; once appointed, such physicians were expected to serve wherever there was a need. In 1888, Congress decreed that officers of the Marine Hospital Service should have titles and pay corresponding to army and navy grades.

This reflected the way in which the services of the Marine Hospital Service were in increased demand: when the Quarantine Act of 1878 was passed, for example, it was the Marine Hospital Service that was expected to oversee it. In the following year Congress did attempt to establish a Board of Health as an alternative mode of dealing with the spread of infectious diseases, but in the face of state-federal conflict and opposition from vested interests (including the Marine Hospital Service), it only lasted four years. Its demise was an indication of the limits of the state's

---

[113] Lacey, 'The world of the bureaus', p. 148
[114] Porter, *Health, civilization, and the state*, p. 147

willingness to incorporate some scientific activities in the central apparatus of government. Some very attenuated attention to medical research
did survive in the form of the Marine Hospital Service's one-room
Laboratory of Hygiene, established in 1887, which conducted research
on infectious disease – the humble beginnings of the National Institutes of
Health.

By 1902 the Marine Hospital Service had expanded to the Public
Health and Marine Hospital Service, and by 1912 it had become simply
the Public Health Service – an indication of the extent to which the
problems of public health were becoming increasingly to be seen as the
core business of the federal government. This transformation was in part
a response to the growing field of bacteriology, which provided a scientific
basis for the need for greater government involvement in dealing with the
spread of disease. This message found fertile ground in fin de siècle
United States, where Progressivism, with its belief in the possibilities of
progress through scientific management, was gaining ground.[115]
Progressivism, with its commitment to applying scientific methods to
cure social ills, also provided a basis for the adoption of eugenic views
about the scientific selection of population, whether through regulation of
marriages or immigration. By 1914, thirty states had legislated to prevent
marriages of those such as the mentally handicapped and the insane, who
were seen as a liability to society. The first state to go as far as using
sterilisation to achieve such ends was Indiana in 1907.[116] Establishing
what was truly 'scientific' could, as these examples indicate, be a political
process that enabled particular understandings of science to be embodied
in legislation.

The period around the end of the nineteenth century was also accompanied by US participation in imperial expansion, especially in the wake of
the Spanish-American war of 1898. Though national identity was strongly
anti-imperial, this being seen as part of the Revolution which created the
United States, the country was pulled into an imperial role as it emerged as
one of the Great Powers. Like other European countries, it pointed to the
spread of science as a way of justifying its imperial expansion. After the
conquest of the Philippines, a Bureau of Science was established there in
1901. Its research covered key areas of state functioning such as weights
and measures and, significantly, public health. Its establishment, too,
indicated the extent to which the US state now accepted that science was
a normal part of the workings of government.[117]

[115] D. Porter, *Health, civilization, and the state*, pp. 156–60
[116] D. Porter, *Health, civilization, and the state*, p. 169   [117] Dupree, *Science*, p. 293

Such acceptance by government of a role for science was, however, largely based on the utilitarian benefits that science could confer. There was limited room for government-sponsored research. Around the beginning of the twentieth century, however, there emerged in the United States a new pattern of patronage that involved neither the state nor the university: the philanthropic foundations established by some of the tycoons of business. Their role was almost analogous to the aristocracy in providing support for science without the constraints imposed by the state – a late example being the role of the Duke of Devonshire in establishing the Cavendish Laboratory at Cambridge University in 1871. The first such foundation, the Rockefeller Institute for Medical Research, founded in 1901, drew on the example of the Pasteur Institute in France (1888) and the Robert Koch Institute in Germany (1891). Appropriately, the chosen field was one in which the state was less than fully active. There followed in 1913 the Rockefeller Foundation, which supported education, including medical education, across the world: among its beneficiaries were the London School of Hygiene and Tropical Medicine, the Johns Hopkins School of Public Health, the Harvard School of Public Health and the China Medical School. Its international credentials were apparent in its foundation in 1913 of the International Health Commission and the China Medical Board in 1914. Another such foundation was the Carnegie Institute, established in Washington in 1902. Its aim was to 'encourage investigation, research, and discovery'[118] in the face of what was described as 'our National Poverty in Science'.[119] By World War I it had developed a distinctive pattern of grants that supported collaborative research, gave funds to individual investigators and covered the costs of the publication of scientific publications.[120] Like the Rockefeller Foundation, it was a supporter of eugenics research until the outbreak of World War II.

By the time of World War I, science had been firmly established as part of the normal functioning of a US federal government that was expanding its activities in the wake of the Civil War. Government science was, however, very largely applied science, with the state playing only a small part in original research. This was a situation that could be justified, however, by the rise of two other possible patrons of science. The first of these were the universities, the research capabilities of which had expanded with the establishment of graduate schools supposedly modelled on those of Germany.[121] Secondly, there were the private

[118] Kevles, *The physicists*, p. 69
[119] Cochrane, *The National Academy of Sciences*, p. 118   [120] Dupree, *Science*, p. 298
[121] Shils, 'The order of learning', pp. 159–95

foundations, of which the Rockefeller and the Carnegie were the best known. To some extent such foundations spared the state expenditure that might have been considered its natural province, such as research into public health. Such delegation, however, fitted well with a state the underlying philosophy of which was a suspicion of centralisation of power. Science might form part of the republican ideology of the United States with its appeal to rationality and rejection of hierarchy, but the concept of the federal state acting as the primary patron of science had only been partly realised by the outbreak of World War I.

Russia

As it had been since the days of Peter the Great, science in Tsarist Russia was associated with modernisation. In the period from the mid-nineteenth century to World War I, the need for such modernisation was manifest, particularly in the face of rapid industrial expansion. Two momentous defeats in this period, in the Crimean War of 1853–6 and in the Russo-Japanese War of 1905, further underlined the price of failure to modernise. But in Tsarist eyes, science was a two-edged sword: its practical advantages were valued but some of the values associated with it were suspect. For a regime based so solidly on tradition and the sanction of an established church, science, with its emphasis on reason and evidence, was not a natural ideological ally.[122] Yet, for all this ambivalence, science in Russia was more centrally a state-sponsored activity than in most countries. This was a tradition that harked back to the establishment of the St Petersburg Academy of Sciences in 1724, an institution which from 1806 to 1917 was known as the Imperial Saint Petersburg Academy of Sciences – thus emphasising still further the connections with the tsarist regime.

By the late nineteenth century, the Academy had been largely Russified, no longer consisting largely of foreign scientists (and particularly Germans) as it had for about a century and a half after it was founded. It also included the humanities and the social scientists, and their ranks were expanding: of the seventy academicians elected from 1890 to 1917, only twenty-eight were from the sciences.[123] The role of the Academy had not, however, fundamentally changed. It provided the government with advice and promoted science (in the widest sense of organised knowledge), and particularly the abstract sciences through its researches.[124] Even in the late nineteenth century there was still much of the Russian territories to explore

---

[122] Vucinich, *Science in Russian culture*, p. 11
[123] Vucinich, *Science in Russian culture*, p. 214
[124] Vucinich, *Science in Russian culture*, pp. 215, 219

and record scientifically, and the Academy continued to play a major role in this national task. This was reflected in the pre-eminence of exploration-related sciences such as geology, botany, zoology and ethnology within Russia.[125]

Tradition loomed large in the Academy, and it was slow to respond to the development of new disciplines since it largely replaced a vacancy in a particular field with another representative of the same field.[126] National politics also played a part in the election of academicians: one of the most distinguished of Russian scientists of the period, Dimitri Mendeleev, who devised the periodic table, was not elected to the Academy, possibly on account of his liberal politics.[127] The Tsar himself vetoed honorary membership of Maxim Gorky on political grounds. In general, academicians were more conservative in character than university professors.[128]

The view that the universities were hotbeds of radicalism had made the Tsarist regime wary of their expansion, but in the late nineteenth century they had sufficiently increased in numbers and research prestige to challenge the Academy's traditional dominance. Russia's first university had been the University of Moscow (founded in 1755), which was followed by the University of St Petersburg in 1819 – the latter doing something to provide a bridge between the Academy and higher education. By 1914 Russia had ten universities.[129] The reforms that followed the Crimean War in the 1860s led to an expansion in the university sector to the point where, for the first time, it challenged the dominance of the Academy.[130] The assassination of Tsar Alexander II in 1881 was followed by a reaction that resulted in 1884 in a highly restrictive decree that sought to control the universities and direct them to the maintenance of the absolutist regime.[131] Nonetheless, institutional ties between the universities and the Academy continued to strengthen in the period from 1884 to 1917.[132]

Academy and universities apart, there were few other centres of scientific research apart from the Central Board of Weights and Measures, an instrumentality of the state.[133] Mendeleev was director of this institution from 1893 to 1907 and made this position a central one within the Tsarist governmental apparatus by acting as a science advisor. His proposals included measures for the systematic expansion of the economic

[125] Graham, *The Soviet Academy of Sciences*, p. 12
[126] Vucinich, *Science in Russian culture*, p. 214   [127] Graham, *Science in Russia*, p. 45
[128] Graham, *The Soviet Academy of Sciences*, pp. 20–1
[129] Graham, *Science in Russia*, p. 80
[130] Graham, *The Soviet Academy of Sciences*, p. 54
[131] Vucinich, *Science in Russian culture*, p. 103
[132] Vucinich, *Science in Russian culture*, p. 215   [133] Josephson, *Physics and politics*, p. 19

resources of the empire.[134] Another state instrumentality which played a role in the spread of modern technology was the armed forces, which provided access to that symbol of modernity, electricity. The army and navy developed modes of utilising electricity efficiently that were then applied to civilian use in the 1890s.[135]

Science, then, was integrated into at least some sections of the state apparatus, which included the Academy and, less securely, the universities. As the pressure for modernisation gathered pace, the Tsarist regime around the end of the nineteenth century established higher technical schools[136] along the German model. Overall, however, the Tsarist regime's commitment to the advancement of science was a very qualified one that fell well behind that of Germany, Britain and the USA. Science was too uncertain an ally for a tradition-bound absolutist state to be fully embraced.

In a period when the remit of the state was expanding, the role of science within the apparatus of government was expanding with it. A larger population, together with the social problems caused by industrialisation, prompted governments to become more involved in remedying the living conditions of their people. This went so far as the creation of a welfare state in Germany and something approximating it in Britain. Even where a welfare state did not emerge, as in France and the United States, there was a greater concern with accumulating data with scientific veracity to provide the basis for improvement. Expanding industry brought with it problems of urbanisation, but it also brought greater prosperity, which governments sought to foster in competition with other states. The need to provide the ever more sophisticated forms of industry with a reliable scientific base prompted Germany, followed by Britain, France and the USA to establish national standards laboratories. Since such laboratories were particularly preoccupied with the electrical industries, they also provided natural centres for research in physics.

Internal expansion into little-known territories remained important to both Russia and the United States, providing government support for forms of natural history including anthropology. External expansion into empire provided governments with an incentive to invest in major institutions, such as botanical gardens to facilitate 'imperial botany' and the moving of plants around the world, the better to suit the economic goals of the imperial power. Science, then, was becoming intertwined with the structures of government though with a strong emphasis on practical returns. The idea of governments supporting research of a less immediately

[134] Graham, *Science in Russia*, p. 45   [135] Coopersmith, 'The role of the military', p. 291
[136] Josephson, *Physics and politics*, p. 17

directed kind, however, was beginning to become a reality with the German initiative of the KW Society in 1911 and the French of the Caisse des recherches scientifiques (Scientific Research Fund) in 1901. Across Europe, universities were becoming more involved in scientific research, often with the aid of governments or (in the United States particularly) philanthropic foundations. Science was becoming more closely integrated into state structures that were soon to be tested to the full by two catastrophic wars.

7   From War to War, 1914–1945

Introduction

The development of the state and the development of war were closely intertwined. A major incentive for the formation of the state was to facilitate control over the territory and resources needed to wage war. As the state grew in size and potency, it could devote more resources to acquiring more complex weapons that drew on a process of experimentation which paralleled that of the sciences. But, on the whole, science as a body of theory had only limited immediate impact on the conduct of war. With the coming of the total wars of the twentieth century, the resources of the state were more fully devoted to the war effort, drawing science more closely into the waging of war. This required a major shift in the state's relations with science: where traditionally science had sought as much independence as possible, the needs of war meant a much closer connection with the state. Combining some element of scientific independence and initiative with direction by the state and fruitful use of its resources required new thinking and new institutions. Devising such institutions in the midst of the flurry of war was no early task, and they were sometimes slow to emerge. Some lessons were drawn from World War I, but World War II was also to involve unique demands on both the state and the scientific estate.

Both wars were very much wars of industrial production, drawing on the vastly increased productive capacities of the nations that had incorporated the Second Industrial Revolution into their economic life. This brought with it a scientification of industrial production, particularly of the leading sectors of the chemical, electrical and metallurgical industries. Science, then, by the outbreak of World War I was increasingly a part of the economic life of most of the major combatants in the war, and the war was to make it even more so. Like an electrical dynamo, war speeded up the pace of change in almost every compartment of life, including science and technology. The development of early forms of technology like the radio or the plane was accelerated by the imperatives of war. Beating plough shears into swords, the war turned into weapons such recent procedures as obtaining nitrogen from the air through the Haber process for making ammonia – essential for Germany's manufacture of explosives. Modern warfare and modern science proved natural partners to a degree that was to change the peace as well as war.

The United States

World War I

Removed from the perennial conflicts between the European states, the United States was slow to enter World War I. Gradually, however, it was compelled to realise the extent to which the war had spilled beyond the borders of Europe and was incorporating whole sectors of the globe, including US trade routes, into the zone of belligerency. Reluctant to admit the prospect of war, the United States was slow to undertake the reorganisation of its scientific and industrial resources that would be required for waging war. It could, however, draw on the experience of some of its major firms in supplying war materiel to the Allies. Such firms, in turn, drew on industrial laboratories that had become increasingly common in the United States from the beginning of the century[1] as technology became more scientific in character. Appropriately, the model for such firms was Germany, the nation where science and technology had first been systematically interlinked. When the United States finally went to war in 1917, then, it had considerable industrial experience in producing armaments.

What it had less experience in was organising scientific research to assist the war effort, and the short duration of the US involvement in the war meant that there was only a limited opportunity to develop appropriate new institutions to fill this niche.[2] In response to the early experiments with manned flight, it had established the National Advisory Committee for Aeronautics in March 1915, for 'the scientific study of the problems of flight with a view to their practical solution'.[3] This Committee was innovative not only in the form of technology to which it was devoted but also in the way it was constituted. Members were drawn from both government and nongovernment backgrounds and it was supplied by Congress with a budget which could be used to foster research in both federal and non-federal institutions. Though its budget

[1] Reich, *The making of American industrial research*, p. 249
[2] Steen, 'Technical expertise', p. 103   [3] Roland, 'Science and war', p. 263

was small, the model allowing a certain degree of independence was to prove influential.[4] It was an important step along the way of the United States devising forms of science policy that enabled it to combine two principles which were in tension: that war promoted centralisation, and that the US constitution sought to avoid centralisation of power. The eventual solution to this conundrum was to outsource much of the scientific research to institutions such as major firms or universities that were distinct from government but could be funded to undertake government-sanctioned research.

Within the armed forces it was the navy which was most interested in innovation in the face of looming war. Following the sinking of the *Lusitania* by a German submarine in 1915, with loss of life which included 120 American passengers on board, the US Navy established a Naval Consulting Board.[5] The problem to which it was particularly devoted was combating submarine warfare, which devastated much of the US trade. This prompted new forms of scientifically informed military technology such as sonar detection devices. The natural instinct of the navy and the army in dealing with experts was to make them part of the armed forces, complete with uniform[6] and lines of command, which, inevitably, limited their freedom of action. Under Thomas Edison the Naval Consulting Board tended to rely on advice from engineers rather than scientists, which prompted the scientists to found another institution.[7]

This was the National Research Council, which was founded as part of a reform program for the National Academy of Sciences. Established in 1916, the National Research Council identified key problems and then parcelled these out to universities and government agencies. This was a major step in the direction of decentralising research away from government, but its effectiveness was limited by the fact that the National Research Council lacked a congressionally sanctioned budget (though it did receive grants from the Rockefeller and Carnegie Foundations).[8] One of the areas of munitions in which the Council promoted research was in the production of poisonous gases. Here it worked with a major government agency, the Department of Mines, as well as farming out problems to a wide range of universities.[9] The National Research Council was valued enough to be made a permanent part of the National Academy with presidential encouragement in May 1918; it was not, however, valued enough to be granted a congressional budget. Its income between

[4] Greenberg, *The politics*, p. 59   [5] Agar, *Science in the twentieth century*, p. 97
[6] Roland, 'Science, technology and war', p. 563
[7] Zachary, *Endless frontier, Vannevar Bush*, p. 36
[8] Pattison, 'Munitions Inventions Department', p. 550
[9] Steen, 'Technical expertise and U.S. mobilization', p. 117

the wars was to continue to come from private sources, principally the Carnegie and Rockefeller Foundations.[10] War with Germany meant not only the production of yet more munitions, but also the substitution by local manufacturers of commodities previously imported from Germany, such as optical instruments, rubber, pharmaceuticals and key chemicals.[11] Among the National Research Council's tasks, then, was to attempt to coordinate domestic production of such goods. Replacement of optical glass was, for example, delegated to the Bureau of Standards and the Geophysical Laboratory of the Carnegie Institution.[12]

World War I made evident the extent to which a modern economy depended on scientifically informed technology. The traditional American confidence that the necessary technology could be produced by private enterprise or the inspired lone inventor such as Thomas Edison was challenged by the enormity and variety of the scale of production. War called for rapid answers, whether to submarine warfare through sonar devices or to the dominance of the machine gun with the tank. Developing such devices could not be left to the vagaries of the market, but required some central direction of the kind provided by the National Research Council. For all the engrained opposition to centralisation, the war had shown that certain forms of central direction could be combined with decentralised production. These were lessons which were to be built on in World War II. The conduct of World War I also left its mark on the peace that followed, as Hunter Dupree argues.[13] War had shown the utility of research, and this lesson was absorbed by much of the industrial sector. It had also shown the advantages of large-scale cooperative research, and this too was a lesson passed on to peacetime America.

Between the Wars

On the return of peace, the cry was for 'normalcy', a reversion to the way things were before the war. Where possible the powers of the federal government were stripped back without the accretions it had gained during the war. This was not an environment to encourage expansion of the federal government's role in the promotion of science – nonetheless, some of the wartime experience began to strengthen the ties between the central government and the scientific enterprise. One example was the way in which the experience of World War I prompted the navy to found

[10] Smith, *American science policy*, p. 30
[11] Reich, *The making of American industrial research*, pp. 253–4
[12] Cochrane, *The National Academy of Sciences*, p. 322   [13] Dupree, *Science*, p. 323

the Naval Research Laboratory, which, along with more directly naval projects, supported fundamental research such as the physics of the Earth's atmosphere.[14]

Having been made permanent in 1918, the National Research Council remained in being acting as a sort of clearing house for scientific projects by farming them out to appropriate university or industrial specialists. Another major function of the Council was to administer grants from the Carnegie and Rockefeller foundations that made possible such programs as the Council's research fellowships.[15] Indeed, in the period between the wars, the Rockefeller and Carnegie foundations continued to do much to promote scientific and medical research in a way that was largely unique to the United States. Between them, these two foundations provided 85–90 per cent of foundation financial support for science in the interwar period.[16] In this period, too, they widened their criteria for grant holders to include individuals (usually located in universities) rather than, as hitherto, large-scale institutions.[17] Nowhere else did foundations act on such a scale.

Their prominence in the support of research meant that these foundations could shape the nature of the science which was researched: the Rockefeller Foundation, for example, doing much to promote molecular biology.[18] Both within the United States and abroad the Foundation's choice of projects to be funded encouraged the dominance of laboratory-based projects over field biology.[19] The presence of the foundations solved some of the anxieties of both government and the scientific estate: the former that the demands of science would drain its coffers and expand its remit unduly, the latter that government money would bring with it government control.[20] Government hoped, too, that it might be possible to reduce further the calls for federal support for science by funds from industry, which, following the lessons learnt in World War I, was more involved in combining research with manufacture. The result was the short-lived National Research Fund, established in 1926 to encourage industry to provide funds for basic research. Such funds were, however, largely unforthcoming, and the result was its disbandment in 1934.[21]

By the 1930s American science was of world standard, one telling index being that the United States produced one-third of the world's papers, a figure that it has continued to reach.[22] A good deal of the credit for this belongs to the foundations. The role of state governments in supporting

---

[14] Hevly, 'The tools of science', pp. 215–16   [15] Dupree, *Science*, p. 329
[16] Kevles, *The physicists*, p. 3   [17] Stanley Coben, 'American foundations', p. 232
[18] Agar, *Science in the twentieth century*, p. 171   [19] Harwood, 'Universities', p. 105
[20] McGrath, *Scientists, business, and the state*, p. 55   [21] Dupree, *Science*, p. 342
[22] Rose, *Science and society*, p. 200

state universities that served to decentralise the scientific enterprise also merits recognition. Ultimately, however, even the foundations had their limits, particularly as the Rockefeller Foundation was involved in other causes such as the promotion of education and medical research internationally. By the 1930s, too, the Depression was beginning to bite, reducing the funds available to the foundations. Science was changing, with the 1930s seeing the emergence of 'Big Science' requiring large and expensive facilities, a trend that was to be greatly heightened by World War II. The most outstanding example of 'Big Science' in physics in the 1930s was the foundation by Ernest Lawrence in 1931 of the Radiation Laboratory at the Berkeley campus of the University of California. This grew so substantially in size and influence as to be eventually renamed the Lawrence Berkeley National Laboratory. Governments, both state and federal, were to be major donors, though so too were the foundations. The federal government met 22 per cent of the costs of its cyclotron before 1940, with the state of California contributing 40 per cent and the foundations 38 per cent.[23]

The need for government to become more involved became more pressing with the downturn of the 1930s. The hope that science could provide means of dealing with the Depression, as it had with wartime needs, prompted the formation of a short-lived experiment, the Science Advisory Board, which lasted from 1933 to 1935. This came about when it was proposed to reorganise the Weather Bureau in the light of recommendations from the National Research Council. What the Council recommended, however, was a more root-and-branch reform with a general board to deal with any such issues where government and science intersected – the proposed board being answerable to the National Academy of Sciences and the Council. As an experiment in scientific organisation, it was given seed money by the Rockefeller foundation but later called on the resources of government. One of its aims, for example, was to facilitate the employment of scientists through the New Deal's Works Progress Administration. The ultimate goal of the Science Advisory Board was to provide a centralised reference point for both scientists and government[24] – a structure which they hoped would avoid the subordination of scientists to the military evident during World War I.[25] Unlike the Academy and the National Research Council, which were separate from government, the Board was to be an integral part of the state's bureaucratic structures. Such a role did not, however, come to pass, since both sides of this putative partnership had reservations about the role of the state in fostering

[23] Seidel, 'The origins', p. 28   [24] Dupree, *Science*, pp. 350, 361
[25] Greenberg, *Politics of American science*, p. 113

science.[26] These were not altogether allayed by the Board's aspirations to disperse control through the network of grant-receiving universities and other institutions rather than centralising it under the president.[27] Even elements of the Academy were unwilling to support a continuation of the Board for fear that government support for science would weaken their autonomy – especially in the highly charged political atmosphere of the New Deal.[28]

The Board, then, expired in 1935, but its efforts were not entirely in vain. It helped to create the climate of opinion for an increase in federal funds for scientific research – from 1935 to 1936 government funds to support research increased by 33 per cent.[29] When World War II loomed, it was very much the same circle of scientific advisers as had been associated with the Science Advisory Board that the President called upon to plan American scientific readiness for war. The contacts made and experience of government of the Science Advisory Board were to yield ample fruit. Another failed attempt to give some central government direction to scientific research was made by the National Bureau of Standards. This, however, foundered on reservations about giving the federal government too much power in this area. Again, however, many of the proposals came to pass during World War II.[30]

Congress was most responsive to requests for research in the medical sciences, which was seen as being particularly valued by the general population. The emerging partnership between the federal government and scientific research received an early manifestation with the creation of the National Cancer Institute in 1937.[31] Medical research, and particularly epidemiology, involved a degree of monitoring of the population that made the federal government more of a 'data state', with the controls this brought over the citizenry. Progressivist ideas about the need to order society along rational lines still influenced thinking in the period up to World War II. Confidence in the ability of science to provide criteria for selection had been evident in the way in which during World War I America had been at the fore in using psycho-physiological tests when choosing recruits, and especially those for the air force.[32] Another more widespread manifestation of this rationalising spirit was the continued popularity of eugenics. Between 1900 and 1940, over two-thirds of American states introduced sterilisation laws with the support of the

[26] Auerbach, 'Scientists in the New Deal', p. 457
[27] Genuth, 'Groping towards science policy', p. 266; Pursell, 'The anatomy'
[28] Cochrane, 'The National Academy of Sciences', p. 363
[29] Auerbach, 'Scientists in the New Deal', p. 480
[30] Pursell, 'A preface', pp. 145–6, 164   [31] Greenberg, *The politics of pure science*, p. 59
[32] Rasmussen, 'Science and technology', p. 310

medical profession. Such attitudes were also reflected in the restricted immigration policies of the 1920s.[33]

There was, then, widespread support for science and its methods, and this had been heightened by the contribution that scientific research had made to the war. Industry, too, was increasingly open to the need for research laboratories to produce goods that linked science and technology. Translating such support into congressional dollars was, however, subject to a number of major obstacles. First and foremost, both scientists and politicians were wary of giving too much power to the state in this area: scientists feared political interference, and the state feared increased expenditure and an augmentation of federal power that they considered inconsistent with American traditions. In political terms, too, while the electorate may have valued science in the abstract, they often needed further persuasion that it could change their lives before being willing to expend tax dollars on its promotion. World War I had lessened such objections, but not to the point where a secure alliance could be formed between science and the state. World War II and its aftermath were, however, to consummate such a linkage with consequences that still shape our world.

World War II

In the prelude to the American entry into World War II the United States had the advantage of that most valuable of commodities, time, to organise its scientific resources.[34] Partly as a result of the work of the National Research Council and the Science Advisory Board, it was recognised in government circles that the impending war was likely to be fought with scientific as well as traditionally military expertise. There was, then, presidential support for the establishment of the National Defence Research Committee in 1940 under the chairmanship of Vannevar Bush, then president of the Carnegie Institute. This was a new structure reflecting a new situation for both the US scientific estate and the state. The probable war, it was clear, would require the mobilisation of the nation's scientific resources, reflecting the centralising trends endemic to war. Yet, regimentation of science ran counter to its own way of doing things, as well as to the decentralising federal form of US government. With considerable success, the structures which Vannevar Bush created – the National Defence Research Committee (NDRC) in 1940 and the Office of Scientific Research and Development (OSRD) in 1941 – overcame such reservations and set the stage for a quite different relationship

---

[33] Largent, *Breeding contempt*, pp. 8, 38, 55, 64   [34] Kevles, *The physicists*, p. 334

between science and the state in the postwar period. The innovative nature of these institutions is underlined by Guy Hartcup's description of the NDRC as 'the first civilian scientific organization specifically designed for war'.[35] Most particularly, the NDRC and the OSRD combined two principles that had hitherto been thought incompatible: government funding for science and a fair measure of independence on the part of the researchers. The main instrument that Bush used to square this circle was the research contract. This allowed for government funding while permitting the research to be carried on not in some centralised government laboratory, but in a university or industrial setting.

7.1 Prof. Vannevar Bush of MIT testifying during the Johnson hearings.

---

[35] Hartcup, *Effect of science*, p. 8

It might have been expected that Bush would draw on an existing institution such as the National Research Council as the foundation for wartime mobilisation of science. The National Research Council, after all, had been the vehicle for such mobilisation in the previous war. The scientists' memories of the last war, however, had generally not been happy ones: the organisations in place did not protect them from a large measure of control by the military, nor did they facilitate direct expenditure of state funds on a particular project. Bush, then, turned to another institution with which he was familiar, the National Advisory Committee on Aeronautics, of which he had been chairman since 1938 and of which he was to remain a member until 1948. The genius of the way the National Advisory Committee on Aeronautics was structured when established in 1915 was that, like the NDRC and OSRD, it combined the status of a government instrumentality with its own congressional appropriation and the power to make contracts with external bodies for undertaking research.[36]

The *modus operandi* of the agencies that Bush created was that of the National Advisory Committee on Aeronautics: projects were contracted out to particular institutions that were much better placed than individuals to resist government centralising pressures. The reciprocal nature of the agreement was given legal form with a formal contract. By war's end the OSRD had issued contracts to 2,300 private agencies with a value of half a billion dollars.[37] Particular institutions stood out as centres of research and as beneficiaries of the OSRD. Research on the self-detonating proximity fuse was originally carried out under the terms of an agreement between the NDRC/OSRD at the Carnegie Institution. In 1942, the host institution was switched to Johns Hopkins, stimulating the development there of an Applied Physics Laboratory. The development of radar was the work of the Radiation Laboratory at MIT; rocket propulsion that of the California Institute of Technology; while the development of the atomic bomb was spread among the universities of California, Chicago and Columbia.[38] As these instances suggest, though the whole strategy of the OSRD was to prevent centralisation in the hands of government, there was a degree of centralisation of research in particular elite institutions.[39]

Nonetheless, the overall outcome of Bush's approach to counterbalancing the centrifugal forces of government was, as Don Price puts it, that he constructed a 'new form of federalism'.[40] It was a form of federalism that

[36] Cochrane, *The National Academy of Sciences*, p. 391; Kevles, *The physicists*, p. 293
[37] Owens, 'Science in the United States', p. 828
[38] Greenberg, *The politics of pure science*, p. 84; Abelson, 'Knowledge and power', p. 472
[39] Penrick et al., *The politics of American science*, p. 46
[40] Price, 'The scientific establishment', p. 33

allowed most scientists to continue working in their own familiar milieu, whether university or industrial laboratory, but also allowed considerable contact with the central government by those directing the OSRD. Hence, Bush became effectively Roosevelt's chief scientific adviser. Such a liaison helps to account for the OSRD retaining its wide brief when other agencies sought some of the power which it had accrued.[41] Another indication of its ability to enlist government support was the success of its measures to ensure that scientists needed for research were not conscripted into the armed services: out of the 10,000 men for whom it sought deferral only 64 were refused.[42]

This is not to say that other existing institutions did not continue to play a role in the scientific mobilisation for war. The work of the military's own technical bureaus continued, being supplemented by the OSRD. So, too, did that of the National Advisory Committee for Aeronautics.[43] Bush discouraged the possibility of the Academy assuming leadership of the scientific war effort[44] on the grounds that its traditions and structures would not facilitate a close relationship with government. Nonetheless, the National Academy and its affiliated National Research Council were a regular source of advice for the OSRD.[45] The National Bureau of Standards, which, between the wars, aspired to a leadership role in bringing science and government together, also was a source of scientific guidance. When first the uranium issue arose as a source of a possible super weapon in 1939, it was first allocated to a committee headed by Lyman J. Briggs, director of the Bureau. The lukewarm assessment of the Briggs committee was that the uses of uranium as a weapon 'must be regarded only as possibilities'.[46] Subsequently, the issue seemed more pressing, particularly in the light of the British Tizard mission to the United States of 1940 (formally known as the British Technical and Scientific Mission under Henry Tizard). Along with other invaluable information brought by the mission, such as that on the construction of the proximity fuse and improved radar, the United States was given the Frisch-Peierls memo on how an atomic bomb might be made. Hence, in early 1941 the possibility of a bomb was referred to the Academy, which recommended 'a strongly intensified effort' to develop the weapon.[47]

The early development of the bomb fell to the OSRD, which had an overall directing role over US scientific mobilisation for war. This both

---

[41] Pursell, 'Science agencies in World War II', p. 375
[42] Zachary, *Endless frontier, Vannevar Bush*, p. 186   [43] Kevles, *The physicists*, p. 298
[44] Lapp, *The new priesthood*, p. 57
[45] Cochrane, *The National Academy of Sciences*, pp. 406–7
[46] Greenberg, *The politics of American science*, p. 108
[47] Baxter, *Scientists against time*, pp. 424–5

served Bush's purposes and, as an anti-New Deal conservative,[48] aroused his suspicions about the uses to which such a centralising institution might be put. Thus, he was adamant that the OSRD should be regarded as a wartime expedient that must be dismantled at war's end. The umbrella quality of the OSRD meant that it could serve as a flexible holding company for a number of separate entities. At its foundation in 1941 it absorbed the NDRC as one of its constituent elements (the NDRC only identified key areas rather than developing actual prototypes in the manner of the OSRD). Shortly afterwards, the Committee on Medical Research also came under the OSRD's broad wing. In 1943, the OSRD also established another branch, the Field Service, which provided first hand scientific advice in the combat zone – the arrangement based on the difficult but successful cooperation between civilian scientists and the military.[49]

The Manhattan Project (the project to build an atomic bomb) also was under the direction of the OSRD but, given its size, the actual mechanics of the process were delegated to the Army Corps of Engineers. This was a project which required a very high degree of joint endeavour by the military and the scientists, the latter of whom much resented the restrictions placed on them in the name of security. At the key site of Los Alamos, however, the arrangement was made to work, partly because of the scientific leadership of J. Robert Oppenheimer, and because at least some autonomy from the military was acknowledged by making Los Alamos a campus of the University of California.[50] Indeed, the Los Alamos site was run under contract between the University of California and the army.[51]

The explosion of the atomic bomb over Hiroshima on 6 August 1945 changed forever the public perception of the role of science, and hence the relations between science and the state. Science now plainly was capable of changing history and was an essential ally in both peace and war. The mushroom cloud was to hang over postwar discussions about the links between science and the federal government, providing an impulse to draw the relationship closer. Along with the bomb the war had produced other potent weapons: the proximity fuse and sophisticated radar systems, among others. It had also produced such benefits for peace as commercially producible penicillin. Bush summed up the impact of the research sponsored by the OSRD when he wrote in the foreword to an official history of the OSRD that 'World War II was the first war in human

[48] Appel, *Shaping biology*, p. 15
[49] Baxter, *Scientists against time*, p. 126: MacLeod, 'Combat scientists', p. 129
[50] Greenberg, *The politics of American science*, p. 119
[51] Dupré and Lakoff, *Science and the nation*, p. 10

7.2 Los Alamos National Laboratory after World War II.

history to be affected decisively by weapons unknown at the outbreak of hostilities.'[52]

A few weeks before the Hiroshima bomb there appeared in July 1945 what amounted to Bush's manifesto on the relations between science and government: *Science – the endless frontier*. The impact of this work was then magnified by the reaction to the bomb and its awesome power. Though it was not to be fully successful in its recommendations, it provided much of the framework for discussion of the relations between science and government in the postwar period. In particular, its recommendation that there should be a state-supported National Science Foundation to oversee government-supported research was eventually to come to pass – though in a much different and much reduced form than Bush had proposed. Bush did at least achieve his goal of ensuring that after World War II the relations between science and government would not once more largely unravel as they had after World War I.[53] Based around four committee reports on the possibilities of scientific research in relation to medicine, public welfare, scientific education and publication of scientific information, the work

[52] Stewart, *Organizing research*, p. ix   [53] Mukerji, *A fragile power*, p. 53

opened with a fluent synthesis of their findings by Bush. With a view to its public impact, Bush appealed to the US pioneer spirit by invoking the resonant metaphor of science as a frontier – a frontier, moreover, with endless possibilities.

Bush also showed political skill by using his work to argue for the benefits of scientific research for both the scientific community and the general populace. This was partly the outcome of his elastic conception of 'basic research' that could serve the needs of both the advancement of pure research and technological progress.[54] A premise of his work was that there was a linear process of development from basic research to technology, or, as Bush himself put it: 'Today, it is truer than ever that basic research is the pacemaker of technological progress.'[55] Though this assumption was later questioned,[56] at the time it was largely accepted – in large part because of wartime experience. The equation between basic research and technological progress, then, provided much of the political impetus for the closer linkage between science and government that was to be a feature of the postwar United States. The need for government support, argued Bush, was particularly necessary since war-torn Europe could not be relied upon to produce the basic research which was 'scientific capital'. In a deftly political argument, Bush also drew on the wartime experience to argue that 'scientific research is absolutely essential to national security . . . only Government can undertake military research'.[57] Overall, *Science – the endless frontier* provided an eloquent statement of the possibilities that had been realised during the war and that might be achieved in peace. Thus, it stated for another age the Baconian ideal of science being used 'for the relief of man's estate'. In his biography of Bush, Pascal Zachary describes the work as 'a kind of creation myth . . . about the new world conceived by the union of science and government during the war'.[58]

Bush was not alone in seeing that the wartime lessons in the use of the alliance between science and the state might be applied to peace. His chief rival in promoting plans to bring about such an alliance about was Harley Kilgore, a Democratic senator from West Virginia, who had been active in proposals for the organisation of science since 1942. Consistent with his strong support for the Roosevelt New Deal, Kilgore saw science as a means of social improvement. Where Bush viewed science largely through the lens of a scientist, Kilgore saw science as a force for reform and renewal of the larger society. Hence, in contrast to Bush, he placed

[54] Pielke, 'In retrospect', *Nature*, 466, 2010, pp. 922–3
[55] Bush, *Science – the endless frontier*, p. 19   [56] Smith, 'World War II', p. 307
[57] Bush, *Science – the endless frontier*, pp. 6, 17
[58] Zachary, *Endless frontier: Vannevar Bush*, p. 221

considerable emphasis on applied science and the social sciences. For Bush, science was, by its nature, an elite pursuit that should be carried out in institutions chosen on meritocratic grounds. By contrast, Kilgore regarded scientific research as a form of government largesse that should be spread throughout the country and among a variety of institutions apart from universities.[59] The existing model which most influenced Kilgore was the Department of Agriculture with its network of applied research stations across the nation.[60]

Most importantly for the future structure of the National Science Foundation, Kilgore was strongly in favour of democratic controls over the directing committee of the Foundation, with a clear line of command to the president. Bush, by contrast, viewed science as a matter for scientists,[61] with a committee consisting of part-time members and a director who would be elected by the other committee members rather than appointed by the president.[62] As in Bacon's Salomon's House, the scientists claimed a considerable degree of autonomy. Too much direct political control, Bush feared, would imprison science in the government-erected bureaucracy of the New Deal to which the conservative Bush was opposed.[63] But in President Truman's view, Bush was asking for an undemocratic degree of independence for scientists – hence Truman's veto of the 1947 proposal for an NSF on the grounds that:

The proposed National Science Foundation would be divorced from control by the people to an extent that implies a distinct lack of faith in democratic processes . . . Full governmental authority and responsibility would be placed in 24 part-time officers whom the President could not effectively hold responsible for proper administration . . . Neither could the Director be held responsible by the President for he would be the appointee of the Foundation.[64]

Whatever the structure of the governance of the NSF, it was a basic principle of Bush's that scientists should have freedom of action in their research. In his mind, as he told Congress in May 1945, there were

two basic principles of successful Government participation in scientific research. First, the research organization must have direct access to Congress for its funds; second, the work of the research organization must not be subject to control or direction from any operating organization whose responsibilities are not exclusively those of research.[65]

---

[59] Penick et al., *The politics of American science*, p. 129
[60] Smith, 'The United States: the formation', p. 39
[61] Wang, 'Scientists and the problem', *Osiris*, 17, 2002, pp. 323–47
[62] Kevles, 'The National Science Foundation', p. 16; Kevles, *The physicists*, p. 356
[63] Appel, *Shaping biology*, p. 30
[64] Penick et al., *The politics of American science*, pp. 135–6
[65] Cochrane, *The National Academy of Sciences*, p. 435

Neither Kilgore nor Bush was successful in having their conception of the National Science Foundation accepted in 1945. Such an outcome did not come until 1950 with a bill that was cosponsored by Kilgore and Bush. The mode by which the NSF was to be governed was closer to Kilgore's conception than Bush's, with the head of the directing committee answerable directly to the president. On the other hand, the NSF, as it finally was embodied, was largely concerned with basic research in a manner that was well removed from Kilgore's socially inclusive role for science. Both men, however, had to accept that the NSF, when finally constituted, was only a very partial realisation of their vision: it was specifically to deal with basic science, while other aspects of science (including medicine) were dealt with by other agencies. Most importantly, it had lost its function as an overall directing and coordinating body.[66] It was a realisation of the NSF which gave little comfort to another of Bush's critics, Henry A. Wallace of the National Bureau of Standards, who had seen *Science – the endless frontier* as drawing too marked a line between academic and government science. Wallace had hoped for greater integration between the NSF and the existing government bureaus dealing with science, and especially, of course, with the National Bureau of Standards.[67]

World War II concluded with a variety of visions about the way in which science and the state could work together. All were agreed that the partnership had worked well during the war and was a major contributor to victory. How to perpetuate this alliance effectively in peacetime was, however, a matter of contention. The major issue was control: how far should receipt of government money determine directions of research. To those, like Kilgore, who envisaged science as part of a more general state-directed improvement of society, government control along the lines of the New Deal was an acceptable price to pay. For Bush, it threatened the focus on pure research and its cultivation by experts that he saw as the heart of any form of National Science Foundation. Scientists had been restive about wartime controls and would be unlikely to tolerate them in peacetime. To a large extent, the regime developed by the NSF, with refereed research proposals accessed on meritocratic principles and then realised in the form of research contracts, assuaged fears about excessive government control. But such procedures were for a different NSF than Bush or Kilgore had envisaged. Eventually, as will be discussed in the Epilogue, other agencies and their claims on government undermined any overall Department of Science-style role for the NSF. But, though

[66] Smith, 'The United States: the formation', p. 39; Smith, *American science policy*, p. 160
[67] Lassman, 'Government science in postwar America', p. 46

Bush did not achieve all of his goals, the basic message of his *Science – the endless frontier* – that science was basic to security and progress – fell on fertile ground. War had expanded the reach of the state to the point where government was to be involved in the promotion of science on a scale hitherto unknown. As so often, war proved to be a major catalyst for the strengthening of ties between science and the state. In the postwar United States this alliance formed in wartime was to survive into peacetime on a scale that would reshape the character of both the state and science.

Britain

World War I

Like the United States, Britain was a liberal democracy with political traditions that emphasised the need to limit the reach of the state lest the liberties of the individual be imperilled. The reflection of such a national mentality in the realm of science policy was to leave the cultivation of science as much as possible to private individuals or institutions rather than the state. But, as outlined in Chapter 6, the growing complexity of the state and industry and the fear of foreign (and particularly German) competition had slowly pushed government into incorporating aspects of scientific investigation into its government apparatus. Three particular initiatives stand out in the early twentieth century, all of which were to be prominent centres of wartime research: the National Physical Laboratory (1900), the Imperial College of Science and Technology (1907) and the Medical Research Committee (1913).

Cautiously and rather tentatively, Britain, then, was moving towards a greater involvement of the state in the conduct of science even before World War I. But, as elsewhere, war greatly accelerated change, requiring Britain to seek compromises that would allow a greater role for the state in the conduct of science while endeavouring not to give the state too much power. The heart of the British scientific establishment remained, as it had since the late seventeenth century, the Royal Society. It had traditionally sought to maintain some distance from the state, seeing itself as an independent organisation, though one willing to give government advice on specific issues. It was, then, a considerable change when, within a few months of war breaking out, the Royal Society offered its services to the nation. As Roy MacLeod writes, 'The beginnings of Britain's war of science date from November 1914 when the Royal Society established its War Committee.'[68] The tentacles of this War Committee spread far with

---

[68] MacLeod, 'The "Arsenal"', p. 49

contact with the Admiralty, the War Office and the Board of Trade. In turn, these and other government instrumentalities sent back suggestions about particular areas to be researched. The Royal Society was also to play a role in establishing bodies dedicated to war-related research, such as the Board of Invention and Research and the Department of Scientific and Industrial Research, the membership of which was to include fellows of the Royal Society.[69] But some of the old reserve remained, with the Royal Society refusing publicly to endorse particular recommendations for government and industry while keeping its own deliberations confidential.[70]

The need for scientific advice became very evident early in the war as scientifically-based forms of manufacture from Germany were no longer available. Products that were particularly in demand were synthetic dyes, optical glass, drugs, magnetos and the tungsten used for making specialist steel.[71] Producing these required state intervention in the market in the light of scientific advice. At the outset of the war there were no obvious instrumentalities to take up such problems, so the responsibility for so acting fell on existing departments of government. Industrial change to produce materials no longer forthcoming from Germany was particularly urgent, and managing this fell to the Board of Trade. Issues of scientific research came under the Department of Education, which initially focussed on the reorganisation of British universities and their research output.[72] In July 1915, a committee of the Privy Council was established to consider issues relating to science and industry, and which received expert advice from an Advisory Council. By 1915, too, the Ministry of Munitions was established, which worked to make industry meet the war's demands, a task often requiring scientific advice. One indication of the success of such scientific mobilisation for war was the establishment of British Dyes in 1915, with the government providing half the capital, a break from laissez-faire orthodoxy that would have been unthinkable before the war. Over the course of the war British Dyes' output was to increase some fivefold.[73]

The Advisory Council and the Privy Council committee were increasingly called upon as it became evident that success in war depended on a fusion of science and technology. As a consequence, these advisory bodies evolved into the Department of Scientific and Industrial Research (DSIR), which was established in December 1916. It was a smooth transition made possible by the expertise and precedents built up by the

---

[69] Lehmann, Morselli, *Science and technology*, p. 14
[70] Hartcup, *The war of invention*, p. 23   [71] Richards, 'Great Britain', p. 179
[72] Varcoe, *Organising for science*, p. 10   [73] Alter, *The reluctant patron*, p. 195

Advisory Council during its short but successful existence.[74] The formation of DSIR marked a watershed for relations between science and the British state. Though, in practice, as we shall see, the DSIR was far from controlling all aspects of scientific and industrial research, it was nonetheless a remarkable departure from previous practice. Where once the emphasis had been on science's independence, the needs of war and the increasing scale of scientific research had resulted in an institution that was an integral part of the governmental apparatus. Government and the scientific community were now working in close partnership. In the United States, by contrast, attempts to erect a centralised Department of Science never came to pass. The DSIR's remit was, however, a limited one: in founding the DSIR, the government's conception was that it should be a body which would promote coordination of research and research training, leaving the actual task of manufacture to others.[75]

Contemporaries were well aware of the significance of the DSIR's foundation and sought safeguards in the face of this new alliance between science and the state. In constructing the administrative machinery of the new department, the DSIR drew on the example of the recently founded Medical Research Committee.[76] The goal of this body was to ensure that its central committee should act as a buffer body between the individual researcher and the state. The modus operandi of the DSIR also attempted to moderate the hand of the state by separating administrative from research supervision. Within the British Commonwealth the DSIR served as the prototype for similar bodies in Australia, Canada, India, New Zealand and South Africa.[77]

It also played a role in the United States in the discussions about the foundation of the National Research Council,[78] though, as we have seen, this body did not have a government budget in the manner of the DSIR. Reflecting their common liberal democratic culture, Britain, like the United States, was to employ the device of contracting out much of their research, thereby inhibiting the growth of too large a government instrumentality. Thus, the DSIR farmed out much of its research to universities and firms. It did, however, also have some in-house facilities that grew in number notably with the transference of the National Physical Laboratory to DSIR control in 1918. Characteristically, once again, there was an attempt to prevent too direct a control by government

---

[74] Varcoe, 'Scientists, government', p. 216   [75] Varcoe, *Organising for science*, p. 76
[76] Alter, *The reluctant patron*, p. 204
[77] MacLeod and Andrews, 'Origins of the DSIR', p. 23
[78] The Medical Research Committee on which the DSIR was based also received praise as a possible model for scientific research in Vannevar Bush's *Science – the endless frontier*, p. 57

over the National Physical Laboratory. Employees came under the DSIR, but research policy was determined by a committee of the Royal Society.[79]

One of the DSIR's chief tasks was to encourage industry to invest in production relevant to the needs of war, particularly in areas where there was a need to substitute for German goods. This involved the delicate task of government intervention in private industry, where the tradition of independence was strong. The DSIR sought to do so through encouragement rather coercion. Soon after the DSIR's foundation, related industries were encouraged to form themselves into Trade Research Associations, with the funding provided by the private sector being matched by DSIR. This was possible thanks to a grant of one million pounds voted by Parliament to establish the Imperial Trust for the Encouragement of Scientific and Industrial Research. The large sum was an indication of how far attitudes towards state sponsorship of scientific research had changed under the impact of war. Not surprisingly, however, the outcomes of this initiative were mixed since many companies were not accustomed to research.[80] Moreover, some companies did not want to invest in research that might aid their competitors. Nonetheless, the DSIR's intervention in the industrial workplace bore fruit: by the end of the war, for example, the British chemical industry had overtaken that of Germany.[81]

The DSIR, however, did not monopolise wartime research. To begin with, there was the Medical Research Committee, which devoted much attention to the abundant medical issues that came with the grinding war of attrition. There were also separate research facilities for each of the armed forces, the Department of Invention and Research for the Navy (1915), the Munitions Invention Department (1915) for the Army and the Air Inventions Committee for the Air Force (1917).[82] These involved civilian scientists working closely with those in uniform, and the results were not always happy. Hence, for example, the closure of the Department of Invention and Research in 1917 despite the important work it had done with anti-submarine devices – notably ASDIC (Allied Submarine Detection Investigation Committee). It was, however, replaced by the Department of Experiments and Research.[83] Such service research facilities drew on prewar origins, another indication of the way in which the British response to World War I could take further developments that had

---

[79] Hutchinson, 'Government laboratories', p. 334; Hutchinson, 'Scientists and civil servants', p. 398
[80] Hutchinson, 'Government laboratories', p. 345   [81] MacLeod, 'The "Arsenal"', p. 47
[82] Pattison, 'Munitions Inventions Department', p. 523
[83] MacLeod and Andrews, 'Scientific advice', pp. 18, 22, 29, 34

been underway before the war. Both the prewar Admiralty and the War Office had such facilities,[84] and an Advisory Committee on Aeronautics had been formed in 1909 that provided a stimulus for inventions research more generally.[85]

World War I, then, had proved a new and different sort of war, demanding of its combatants a level of commitment of human and material resources that was unparalleled. Its voracious demands reflected the conduct of war by nations with economies shaped by the scientification of industry. To match the needs of such a war, Britain was obliged to mobilise rapidly its scientific and industrial resources. Doing so brought greater centralisation of the state's power, which ran counter to the traditional way in which science had operated in Britain. Hence the emphasis on administrative forms that endeavoured to limit the hand of the state in the workings of scientific institutions linked to government. The scale and length of the war led to a remarkable level of mobilisation of Britain's scientific resources – by the end of the war at least half of the nation's civilian scientists were devoted to war work.[86] In a relatively short time, scientists used to university research were redeployed successively to war-related tasks. Many of the universities contributed considerably though the Imperial College of Science and Technology stood out particularly.[87] It was of the nature of war, however, that fundamental research was brushed aside in the race to produce effective weapons of war. As peace approached, one issue for government was to consider how far the successful union of science and the state might be continued in the very different conditions of peace.[88]

Between the Wars

The DSIR was a child of wartime centralisation but, come peacetime, it was not only retained but was expanded, taking over control of the Geological Survey and Museum from the Board of Education in 1919 and the Road Experimental Station from the Ministry of Transport in 1933. Its functioning persuaded the scientific community that it was possible to draw closer bonds with government without science losing its initiative and freedom of thought. Such a balancing of the claims of the state and those of science had been earlier achieved by the Medical Research Committee, and its example had been drawn on by the DSIR. It was, however, the DSIR that was the most visible example of a liberal

[84] Rasmussen, 'Science and technology', p. 310,
[85] Lehmann, Morselli, 'Science and technology', p. 14
[86] MacLeod, 'The scientists go to war', p. 42   [87] Hartcup, *War of inventions*, p. 189
[88] Pattison, 'Munitions Inventions Department', p. 522

state's attempt to harness but not hobble science. The DSIR's functioning received commendation in the 1917–18 review of the Committee of
Inquiry into the Machinery of Government chaired by Lord Haldane.
This inquiry enshrined what became known as the Haldane Doctrine as a
basic operating principle for government organisations that impinged on
scientific research. Such research, argued Lord Haldane, should be
supervised by a general committee, and preferably a research council,
rather than by the administrative wing of a department.[89]

The Haldane Doctrine helped shape the character of the institutions
established between the wars that linked science with the state. Belatedly,
following up a suggestion that Haldane had made in another enquiry in
1903, peacetime brought into being the University Grants Commission
in 1919. This finally brought some order to government funding to
universities, which up to this point had been an *ad hoc* collection of
administrative decisions about individual universities.[90] It also provided
securer funding to the relatively neglected university sector, including the
science faculties. True to the spirit of the Haldane Doctrine, expert
academic advice was separated from the administrative and financial
supervision of government, with the commission serving to facilitate
such a division of powers.

In 1920, the Medical Research Committee was raised to the greater
dignity of a Medical Research Council and embodied in its administrative
structures some of the key features of the Haldane Doctrine in separating
administrative and research functions. Indeed, as an incorporated entity
kept separate from the Ministry of Health, it achieved greater autonomy
than the DSIR, which was, after all, a government department. It was to
act as a model for other research councils when they were established after
World War II. Along with the DSIR and the Medical Research Council, a
third council was established to promote scientific research in the form of
the Agricultural Research Council. This was founded in 1931 and again
maintained the spirit of the Haldane doctrine in having separate supervisory bodies for research and administration.[91] This Council may well
have owed something to the successful example of the American
Department of Agriculture.[92] Where before the war Britain had looked
to Germany for models of scientific organisation, after the war it was more
inclined to look to the United States. Along with this government organisation, the British were also great admirers of the industrial research
laboratories erected by major US corporations (institutions which in their
turn owed much to the earlier German example). Indeed, DSIR gave

[89] Ronayne, *Science in government*, p. 15   [90] Hutchinson, 'Origins', p. 586
[91] Rose, *Science and society*, p. 50   [92] Clarke, Sabine, 'Pure science', p. 290

preference for grants to industry over the universities, the latter still being viewed as primarily teaching institutions.[93]

On the whole, then, between the wars the principles and practice of government bodies relating to science stressed the need for maintaining an element of scientific autonomy. There were those, however, who were heretical enough to challenge the Haldane Doctrine. The major figure leading the movement for greater rather than lesser government control over science was J. D. Bernal, professor of physics at Birkbeck College, University of London. Influenced by the Soviet example and reacting against the Depression, Bernal aspired to assimilate science within an overall planned economy. His manifesto was the *Social Function of Science* (1939), in which he lamented the lack of planning in science, or, as he put it, 'the inefficiency, the frustration and the diversion of scientific effort to base ends'. Rather, he urged that science should be used in a Baconian spirit for 'the relief of man's estate', arguing that the current misuse of science led to 'disease, enforced stupidity, misery, thankless toil, and premature death for the great majority, and an anxious, grasping, and futile life for the remainder'.[94] Bernal may have been the most eloquent spokesman for the movement for a closely planned government direction of science, but he was certainly not alone as admiration for the Soviet planned economy gained ground in Britain from the late 1920s. Predictably, a reaction set in, with a Society for Freedom in Science being established in opposition to 'Bernalism'. Its goal was to advocate the view that 'the advancement of knowledge by scientific research has a value as an end in itself'.[95] Another of Bernal's adversaries, J. R. Baker, decried the way in which 'an ugly new god called the state demands worship'.[96]

In the event, Bernal's call for a centralised supervision of science made little headway in Britain. It did, however, do something to focus discussion on the benefits to be obtained from expanding the size of the scientific estate. The need for more expenditure on science had been urged rather ineffectively by the Science Guild since its foundation in 1905, but in 1936 this was absorbed by the British Association of Science, which had always been an advocate for expanding spending on science. The Guild had joined forces with the Association of Scientific Workers, a trade union for scientists, to establish in 1933 the Parliamentary Science Committee, a lobby group to increase government outlays on science.[97] As the 1930s progressed, so too did the agitation to heed the lessons of

[93] Alter, *Reluctant patron*, pp. 200, 211   [94] McGucken, 'Freedom and planning', p. 42
[95] McGucken, 'Freedom and planning', p. 46
[96] McGucken, 'Freedom and planning', p. 53
[97] McGucken, 'Central organisation', p. 33

World War I and use the machinery of government to prepare Britain for the event of another war – a war in which science would certainly be a key factor.

After 1918, the DSIR largely distanced itself from directly militarily-directed research, but it remained aware of the importance of all aspects of the economy if another war should break out.[98] The DSIR kept a watching brief on military research by establishing in 1919 and 1920 research coordinating boards in engineering and scientific fields, with a view to minimising overlap with civilian research and to give support to research that appeared to have civilian uses.[99] The diffusion of science within industry remained one of the DSIR's primary goals. During the war, one route to assisting such a dissemination had been the trade research associations, where different firms in the same industry pooled resources (along with those of the DSIR) to conduct research. These had been of limited success,[100] but a variant scheme using the cooperative principle to disseminate scientific research with the aid of universities and professional associations had rather better fortunes.[101] The need for industrial research became more of an accepted practice in larger firms between the wars, an exemplary case being Imperial Chemical Industries, formed in 1926 from the merger of the government-sponsored British Dyes with other private firms.[102] Thus was the sudden acceleration in the output of British chemical industries during World War I turned to peacetime advantage. There was, of course, always the possibility of such a chemical industry being once again turned to military purposes – as indeed did happen in World War II.

As the likelihood of another war became more and more depressingly probable during the 1930s, more attention was devoted to the bringing together of scientific research with military needs. The focus was very much on the air war, because of the rather exaggerated but, nonetheless, understandable fears about the impact of bombing on the crowded island of Britain. One response was the establishment in 1934 of the Committee for the Scientific Study of the Air Force, which conducted its early experiments at the DSIR's National Physical Laboratory (though later moving them to the Air Ministry). The final success of the programme in producing radar was the product of both civilian and military scientific researchers.[103] By the eve of the war, the Royal Society had taken the initiative of drawing up a list of scientific personnel to avoid the World

---

[98] Edgerton, 'Science and war', p. 940
[99] Hutchinson, 'Scientists as an inferior class', p. 399
[100] Varcoe, *Organising for science*, pp. 22–3
[101] Varcoe, 'Co-operative research associations'
[102] Haber, 'Government intervention', p. 87   [103] Edgerton, 'Science and war', p. 940

War I situation of sending scientists into battle. All in all, the urgent reorganisation of Britain's scientific capabilities that had taken place in World War I was sufficiently preserved and augmented during the inter-war period. This helped to ensure that, when war finally broke out again, the nation's scientific reserves could be rapidly mobilised.

World War II

When World War II broke out, then, the outlines of a national framework for the merging of scientific and military resources were in place. In contrast to the United States, there was not a fundamental reorganisation of science and research in Britain. The basic structure erected during World War I and refined between the wars remained in being and could be adapted to the needs of another, even more scientifically demanding, war. Peacetime institutions could be readily readapted to wartime needs. The DSIR, which had been conceived in wartime and then turned to peaceful uses, was again turned to military ends. In 1939, Cabinet made the crucial decision not to disperse the staff of the DSIR, but to keep it and its units as functioning parts of a wartime research mobilisation.[104] The crucial discussions about the uranium bomb took place first within a committee of the Scientific Study of the Air Force known as the Maud committee, and were then transferred to a committee of the DSIR with the code name 'Tube Alloys'. The DSIR's National Physical Laboratory was the site for experimentation of novel weapons, including the 'bouncing bomb' used on the 'dam busters' raid of 1943.[105]

As in World War I, the Royal Society was intimately involved in wartime planning of science. It was at its insistence that Churchill was persuaded in 1940 to establish the Scientific Advice Committee to the War Cabinet. This included strong representation of the Royal Society itself (president and two secretaries) as well as the heads of the three research councils, the DSIR, the Medical Research Council and the Agricultural Research Council. Rather than rely on this committee, however, Churchill, in his typically maverick way, turned for advice to a single individual, Professor Lindemann (later Lord Cherwell), who, after 1942, sat in Cabinet in the capacity of Paymaster-General.[106] It was an instance of the extent to which the scientific advice given to states can reflect a range of thinking, often with no clear consensus. As David Edgerton remarks: 'World War II was an experts' war, but also a war between experts.'[107] Adjudicating between such experts meant political decisions

[104] Rose, *Science and society*, p. 68   [105] Pyatt, *The National Physical Laboratory*, p. 138
[106] McGucken, 'Royal Society', p. 87   [107] Edgerton, *Britain's war machine*, p. 123

about matters of scientific import; inevitably, partnership with the state, which was heightened by wartime needs, brought with it an intermingling of the scientific and the political.

The British wartime state, then, could call on considerable scientific reserves and did so with a greater element of planning than in World War I. With such planning went an inevitable expansion in bureaucracy.[108] The Department of Labour kept records of scientific manpower that was then deployed as the needs of the war dictated. The dominant wartime movement was towards the increasing scientification of industry to serve military needs. University scientists were often deployed in quite new settings,[109] and Britain did not have the US phenomenon of a great expansion of particular universities or departments within them in response to government research contracts.[110] One novel way in which the British state was to use its scientific capabilities was as a form of diplomacy. As we have seen earlier in this chapter, in September 1940, when Britain was immersed in the Battle of Britain and the United States was not at war, the Tizard mission (The British Technical and Scientific Commission) brought to the United States details of some of the major wartime scientific research projects. These included details of radar research in the form of the cavity magnetron, which allowed the use of microwaves,[111] jet propulsion and, most importantly, a memorandum on the feasibility of the bomb, which helped to persuade the Americans to review the issue and, subsequently, to commit themselves to the project.[112]

From the British point of view, it was hoped that these projects could be more readily completed with American resources than British, which were stretched to their limit by the war. There was also the hope, too, that joint scientific and technical cooperation would help build up a rapport between the United States and Britain – a hope which was eventually realised. Though brought to a successful conclusion, British and American partnership on the bomb was at times an uneasy one. Given that only the United States had the resources and the spaces free from the risk of being bombed, the project was transferred to the United States, though with the inclusion of British scientists already at work on the project. As the project progressed, however, they were 'compartmentalised'[113] – that is, only allowed access to those aspects which immediately concerned their own work. Though the junior partner, Britain was ceded the right to have to give its consent to the use of the

[108] Edgerton, *Britain's war machine*, p. 227   [109] Rose, *Science and society*, p. 68
[110] Edgerton, 'British scientific intellectuals', p. 5
[111] Bowler and Morus, *Making modern science*, p. 470
[112] Pierre, *Nuclear politics*, pp. 15–16   [113] Pierre, *Nuclear politics*, p. 31

bomb. Come the end of the war, however, Britain was denied access to the information needed to build a bomb of their own by the US 1946 McMahon Atomic Energy Act. Many in Britain viewed this as a violation of earlier understandings about the joint use of the research. Britain, then, had to proceed to build its own bomb using the expertise of British scientists involved in the Manhattan Project. The high level of scientific cooperation of Britain and the United States during the war, then, was in peacetime undermined by the strong impetus of individual states to monopolise potent military technology. The alliance between science and the individual state proved more durable than international agreements about the ends to which science might be put.

As in the United States, the extent to which World War II was a scientist's war, with the use of the bomb as its great denouement, enforced in the public consciousness the importance of science and the extent to which it was related to social and political processes. With war's end it was apparent that Britain had used its scientific manpower more effectively than the totalitarian German state,[114] with such achievements as the Bletchley codebreakers, the development of operations research (the use of complex mathematical models to improve decision making), radar and anti-submarine weapons, along with medical breakthroughs such as the early development of penicillin. The alliance between science and the state was now firmly established, though the postwar world would see continuing debates about the form this should take. Should the more centralised planning of the war be continued in some form or should the traditions of a more independent British scientific estate be once more reasserted? One indication of a turning away from centralised planning was the continuation into peacetime of the DSIR, now a survivor of two wars, with its apparatus for separating the scientific from the administrative. This 'in-between' role of the DSIR was to be expanded with the creation of more scientific councils after the war.

France

World War I

At the outbreak of World War I, France faced similar shortages to Britain, especially when it came to German-produced scientific manufactures. Its problems were further compounded by the fact that, from very early in the war, Germany controlled sections of northern France that included some of its industrial centres and sources of raw materials. Overcoming such

---

[114] Rose, *Science and society*, p. 68

difficulties meant reconstructing the economy and integrating scientific advice in both military and industrial contexts – all of which required decisive state action and a willingness to use the centralised powers that were embodied in the post-Napoleonic state structures. Special attention was devoted to the chemical industry on the output of which much of the war effort depended. Rapid progress needed to be made since, when compared to Germany, France had about one-tenth the number of chemists.[115] Before the war this industry had enjoyed a large degree of freedom from state control, but this began to change very early in the war, when, in September 1914, there was established a state-directed Office des produits chimiques et pharmaceutiques (Chemicals and Pharmaceuticals Office). The research needed to make the chemical industries more effective was largely passed on to scientists within the universities and the existing research institutions – the Institut Pasteur, for example, made dyes for both military and civilian uses.[116] The variety of institutions and their frequent reorganisation was given a fundamental order by the way in which they formed part of the machinery of the French state – a state which increasingly was assuming control of the French economy and, with it, the mobilisation of science for war. Government control increased over the duration of the war, so that in its last year virtually all the economy was subject to the will of the state.[117]

The most central institution for providing government support for scientific innovation was the Direction des inventions intéressant la défense nationale (Directorate of Defense Inventions), which was established in November 1915 by the Minister of Education. As its name suggests, one of its primary functions was to examine inventions to determine their utility for war. Its functions, however, were broader than this, and it was part of its remit to promote general scientific mobilisation for war, including partnership between science and both industry and the military. It formed particularly strong bonds with the universities.[118] The increasing importance of the Directorate as a central point for coordinating the scientific war effort was underlined by the way in which it was gradually elevated to the epicentre of French oversight of the war. From the Department of Education, it was moved to the Ministère de l'armement et des fabrications de guerre (Ministry of Armaments and War Constructions) and placed under an undersecretary of state. The following year it was given the verbose but imposing title of Sous-secrétariat d'etat des inventions, des etudes et des expériences

[115] Paul, *From knowledge to power*, p. 321
[116] Chauveau, 'Mobilization and industrial policy'
[117] Chauveau, 'Mobilization and industrial policy', p. 22
[118] Bowler and Morus, *Making modern science*, p. 467

techniques (Undersecretariat of State for Inventions, Research and Experimental Techniques). To emphasise its elevation, it was later transferred in 1917 to the Ministry of War.[119]

Along with the Directorate of Inventions, there were various specialised units the success of which depended on the willingness of the scientists and the military to work together. One such was the highly productive Service des poudres (Explosive Service).[120] Another was the Section technique de l'aéronautique militaire (Technical Section of Military Aeronautics), which grew out of the prewar Institute Aérotechnique at the University of Paris. It built up one of the best-equipped aeronautical laboratories at the military school at Saint-Cyr, with such refinements as a wind tunnel and facilities for testing motors at different temperatures.[121] These were further instances of the French state's ability to create institutions which would bring together science, industry and the military. France's response to the war and the urgent need to modernise its scientific establishments were to use these different agencies, whether existing or newly created, to link together these different constituencies. The fact that the various participants shared common goals made this state-directed redeployment of human resources a relatively successful one. Under the pressure of war, the centralising elements of its polity manifested themselves strongly but with considerable success.

Between the Wars

World War I had seen a mobilisation of France's scientific resources under the leadership of the central state. Much was retained from this successful redeployment, with the fear of German scientific or, particularly from the 1930s, military dominance fortifying the resolve to maintain a research establishment that could be focussed on national needs. Come peace, the Direction des inventions (Directorate of Inventions) moved from the Ministry of War back to the Ministry of Public Instruction, being restructured and renamed in 1922 as the Office national des recherches scientifiques (National Office of Scientific Researches). It also incorporated the 1901 foundation, the Caisse des researches scientifiques (Fund for Scientific Researches)[122] – an example of the continuing trend in French government scientific circles to centralise administration as much as possible. Other World War I scientific offices also survived in mutated forms. The Office des produits chimiques

---

[119] Lehmann, Morselli, *Science and technology*, pp. 11–13
[120] MacLeod and Johnson, 'Introduction' in MacLeod and Johnson (eds.), *Frontline and factory*, p. xvii
[121] Lehmann, Morselli, *Science and technology*, p. 12   [122] Gilpin, *France*, p. 154

(the Office of Chemical Products) gave way to the Commission des etudes et experiences chimiques (The Commission of Chemical Studies and Experiments), which was established in the 1920s in the Ministry of War. There was continuity, too, from war to peace in the aeronautical industry with the reorganisation in 1919 of the Service technique de l'aéronautique (Technical Service of Aeronautics).[123]

The 1930s was a period of frequent restructuring of France's scientific structures, driven by the scientists' own desire for reform along with the state's fear of the increasing German threat.[124] The decade opened with the foundation of Caisse nationale des sciences (National Science Fund). Established in 1930, this was intended to improve the level of basic research. The decade closed, however, with the merger in 1939 of the various government institutions formed for the promotion of science into an institution which has lasted until the present, the Centre national de la recherches scientifique (CNRS, the National Centre for Scientific Research) under the Ministry of National Education. Its constitution subsumed the various purposes that had prompted the foundation of government scientific agencies. Within the Centre were two divisions, with one for applied research and the other for fundamental; the overall goal being to prepare the nation for war.[125] On the surface, the frequent change in the name and the character of the institutions dealing with science might indicate confusion. On the contrary, however, they represented a reforming spirit and a desire for more effective control over the nation's scientific estate.

Successive agencies also incorporated a wider range of goals for the promotion of both fundamental and applied researches until, finally, the CNRS pulled these strands together in one central, ongoing institution. The CNRS was, however, concerned with government scientific agencies, with the result that it formed a research environment largely separate from the universities (though such separation has weakened in more recent times). The traditional binary between government-supported research units and universities as teaching institutions was thus reinforced.[126] Significantly, one of the major impulses which had fed into the chain of institutional evolution that led to the CNRS was the World War I Directorate of Inventions. The CNRS was also formed under the shadow of the impeding World War II. War and science were close allies. Alas, for France, however, World War II was to result in a rapid defeat that left little time to deploy many of the nation's scientific reserves.

---

[123] Lehmann, Morselli, 'Science and technology', pp. 12–13
[124] Shinn, 'The industry, research, and education nexus', p. 143; Paul, *From knowledge to power*, pp. 340–1, 344
[125] Gilpin, *France*, p. 158
[126] Bulmer, 'Knowledge institutionalized', *Minerva*, 40: 189–201, 2002, p. 193

Germany

World War I

At the outbreak of World War I, Germany had a well-developed scientific establishment that was closely connected to the state. The universities were supplemented by the state-sponsored research institutes, such as the Physikalisch-Technische Reichsanstalt (Imperial Physical and Technical Institute) and the KW Society. Scientific techniques of experimentation were more and more being incorporated into the operations of large industrial firms. Industrial processes meant to produce such civilian products as dyes could be turned to military use, such as the production of poisonous gases. The Haber process for fixing nitrogen from the air could be turned from the manufacture of fertilisers to explosives that made possible the continuation of the German war effort in the face of the British naval blockade. Such 'dual use' was to be the foundation of much of the German war production and, later, its adaptation to the needs of peacetime.[127] In such ways, the war brought with it closer integration of science with the military, with which came a greater measure of state centralisation.[128]

Hopes of a rapid victory meant that such scientific assets were not immediately turned to advantage. As the war turned to a stalemate, there was more of an inclination for the military and the German scientific establishment to work together to produce new weapons. The most spectacular example of this was the use of chlorine as a weapon in trench warfare in January 1915. This was the outcome of the cooperation of the KW Institute for Physical and Electrochemistry under Fritz Haber with the major German chemical companies, which produced chlorine as a by-product of dye manufacture. This was an instance of the way in which Haber encouraged the use of this institute to serve as a focus for war-related research that brought together the state and private enterprise. In doing so, he was, as Margit Szöllösi-Janze argues, developing a new form of institution within the larger Kaiser Wilhelm organisation. The object of such an institution was to allow a high level of direction by the state through the military, which provided much of the finance for an initiative that also involved close cooperation with private enterprise.[129] Such an institution was given a more formal and visible form with the creation of the KW Institute for the Science of War Technology on 1 January 1917,

[127] Johnson, 'Technological mobilization', p. 6
[128] Szöllösi-Janze, 'Science and social space', p. 352
[129] Szöllösi-Janze, 'Science and social space', p. 353

at the initiative of Haber and Emil Fischer.[130] The object was to bring the 'Army and Navy into close contact with representatives of natural science and technology'.[131] By this stage of the war, however, shortages of materials brought about by the British naval blockade limited the scientific possibilities of such an institute. Such shortages strengthened the need for a scientific partnership with the state to create such substitutes for imported products as synthetic rubber.

Universities were less prominent in wartime research than the KW Institutes or the Imperial Physical and Technical Institute, but there were pockets of war-related scientific developments. This was the case at the University of Göttingen, where the army and navy worked with scientists on the development of the radio. There was also research into improvements of aircraft.[132] Universities, however, formed part of the network of institutions that worked to make the German economy as self-sufficient as possible. This was an enterprise that continued after the war, familiarising the population with the possibilities of scientific research.[133]

Unlike Britain and France, then, Germany did not radically change the institutional character of its scientific establishment as a result of the war. There was no new equivalent of the British Department of Scientific and Industrial Research or the French Directorate of Inventions. The existing institutions that had seen prewar Germany become a dominant scientific and industrial force worldwide were considered to meet the needs of war as well as peace. There was adaptation within the existing forms, such as the creation of the KW Institute for the Science of War Technology, which was new in the degree to which it was controlled by the military. But even within existing institutions there were changes in character as the dependence on the military and, with it, state funding increased. The German tradition of decentralisation was challenged but not nullified by the war. There remained a respect for institutional autonomy that was to be tested further in the stormy decades ahead.

Weimar Germany

One of the ways in which Weimar Germany sought to offset the bitter taste of defeat was to console itself with the glories of the German scientific tradition. Germany might have lost in war, but it kept alive such spiritual values as the worth of science and of the forms of civilisation associated with it. Science became what the Germans called a 'Macht-

---

[130] Bowler and Morus, *Making modern science*, p. 467
[131] Hartcup, *The war of invention*, p. 35
[132] Lehmann, Morselli, *Science and technology*, pp. 10–11
[133] Szöllösi-Janze, 'Science and social space', p. 354

Ersatz', a substitute for power, in terms of Germany's international standing. Hence the postwar Allied nations' scientific boycott of German science was particularly wounding since it lessened the prominence of Germany's scientific achievements.[134] While, of course, supportive of the state defining itself in terms of its scientific achievement, Germany's scientists sought at the same time to retain control over their institutions. By doing so, they sought to separate the conduct of science from the left-liberal regime that had replaced the Empire and that had little support among the generally conservative scientists.

Yet, the Weimar Republic provided considerable support for science, investing in maintaining the national self-image of being a leader in science. Science was here being used to support a state identity. It was the Weimar government that gave institutional life to the proposal by some members of the Berlin Academy in 1920 for a Notgemeinschaft der Deutschen Wissenschaft (Emergency Fund for German Science)[135] – 'emergency' because of the fear that German scientific excellence would disintegrate if it were not sustained. Some 80 per cent of the funding for the institution came from the Weimar government.[136] Funds for medical research from the Emergency Fund were supplemented by grants from the Rockefeller Foundation for fellowships, though these were to be phased out by 1939.[137] Other, more general changes also favoured a greater prominence on the part of the Weimar state. The Weimar constitution gave the central government a stronger financial position as against the individual states of the federal system together with powers to establish educational and scientific policies.[138]

Thanks in part to the state support channelled through the Emergency Society, the Weimar period became known for its efflorescence of research in physics,[139] an instance of the way science could be shaped by government policy. Government policy also favoured promising individuals with a research scheme that awarded funds to a particular scientist for a particular project rather than, as hitherto, to the directors of an institution.[140] This, however, did not preclude supporting the KW Society, which went from four to thirty institutes in the postwar period. Before the war, the Kaiser Wilhelm Institutes received little more than their land and the salary of the directors from the state, with most of the funding coming from private enterprise. Under the Weimar government,

---

[134] Schoeder-Gudehus, 'The argument', p. 551; Forman, 'Scientific internationalism'
[135] Rasmussen, 'Science and technology', p. 311
[136] Pfetsch, 'Germany: three models', p. 195
[137] Weindling, 'The Rockefeller foundation', p. 128
[138] Forman, 'The financial support', p. 44   [139] Walker, *German national socialism*, p. 7
[140] Pfetsch, 'Germany: three models', p. 195

50 per cent of funding came from the central government and the Prussian state.[141] Receipt of such funds was accompanied by determined opposition from the KW Society to any loss of their independence.[142] For the Society, as for other scientific centres including the universities, such independence made it easier for scientists to plead an apolitical stance in the face of a regime for which they had little sympathy. Generally, however, the high level of government support proved compatible with maintaining the traditional autonomy of scientific institutions.[143]

In the face of economic depression and political upheaval, the Weimar Republic had succeeded in maintaining the substance of the scientific establishment that had been erected before World War I. In some respects, indeed, it had been expanded with, for example, a greater number of Kaiser Wilhelm Institutes. By such devices as the Emergency Fund, the state had kept alive the reputation of German science, thus reaffirming the self-image of the nation as a scientific power. These achievements of the Weimar Republic were, however, to be obscured in the tumultuous years that followed the fall of the Republic and the ascent of Hitler in 1933.

Nazi Rule, 1933–1945

From 1933 to 1945, Germany was ruled by a totalitarian ideology that shaped every branch of life, including the conduct of science. Such ideology was promulgated and enforced by the state using a range of sanctions. While no state speaks with one voice, the Nazi state was particularly polycratic in character with different power centres: the Party, the government (in the sense of the administrative machinery that maintains the running of the state), industry and the army.[144] To these could be added other possible power centres such as the SS.[145] Of these, the Party, as the upholders of ideology, was the primary power centre. Some aspects of Nazi ideology were enforced with greater rigour than others. Anti-Semitism was central and enforced almost uniformly. Within two years of taking power, Hitler had all Jews, together with others who would not conform to the regime, expelled from the universities: some 15–20 per cent of university teachers were dismissed (with considerable variations between universities).[146] Jews were particularly prominent in the newer branches of physics, so the impact on that science of

---

[141] Macrakis, *Surviving the swastika*, p. 33   [142] Macrakis, *Surviving the swastika*, p. 48
[143] Pfetsch, 'Germany: three models', p. 195   [144] Mehrtens, 'The social system', p. 294
[145] Pfetsch, 'Germany: three models', p. 196
[146] Poletschek, 'The invention of Humboldt', p. 44. This is a revisionist estimate that is
     somewhat lower than previous such estimates.

Nazi ideology was especially marked, with one in four physicists being dismissed.[147] The universities came under the civil service regulations, so they were particularly vulnerable, but Jews were eventually driven out of all scientific institutes, including the KW Society, which preserved some measure of autonomy.

Another strongly held belief of Nazism that was to affect the conduct of science was opposition to internationalism.[148] Scientific contact across nations was part of the lifeblood of scientific enquiry but, in Nazi eyes, it weakened the distinctly Teutonic identity of German science, as well as exposing scientists to dangerous political influences. One of the aspects of science that made it somewhat suspect to the Nazis was its international character and links with liberal democracies.[149] Such lack of international scientific contact and stimulation may help to account for the way in which German science, while still strong during the Third Reich, nonetheless fell behind that of its allies by late in the war and after the war. The boost given to German science before 1933, when freer exchange of ideas was possible, had by that time worn thin.[150]

How far did Nazism have a scientific or pseudoscientific basis? Hitler's social Darwinist view of society led some Nazis to claim that Nazism had a basis in biology. One senior Nazi official claimed that Nazism was 'applied biology'.[151] Hitler himself described the impact of Nazism as 'the final step in the overcoming of historicism and the recognition of purely biological values'.[152] Biology was favoured in the bestowal of grants[153] – an instance of the way in which the state can shape the profile of scientific research. Some of this support came from the SS Ancestral Foundation to support Nazi 'racial science'.[154] Most, however, was of an orthodox kind – illustrating how scientific work of a high standard could be pursued under the Nazis. Biologists were not under pressure to pursue a form of Nazi biology. Indeed, there was no prescribed form of Nazi biology to be enforced.[155]

There was an attempt to argue that some forms of physics were more compatible with Nazism than others. The leaders of the Deutsche Physik (German Physics) movement, Johannes Stark and Phillip Leonard, contended that physics had been corrupted by the influence of Jews (with Einstein the prime suspect). These, it was alleged, had introduced

---

[147] Beyerchen, *Science under Hitler*, p. 40
[148] Schroeder-Gudehaus, 'Nationalism and internationalism', p. 916
[149] Weindling, *Health, race and German politics*, p. 491
[150] Macrakis, *Surviving the swastika*, p. 199   [151] Weale, *Science and the swastika*, p. 19
[152] Proctor, *Racial hygiene*, p. 64   [153] Deichmann, *Biologists under Hitler*, p. 319
[154] Walker, *Nazi science*, p. 51; Walker, 'Twentieth century German science', p. 802
[155] Deichmann, *Biologists under Hitler*, pp. 319, 322

mystical concepts like relativity and quantum mechanics, which obscured the simplicity and experimentally verifiable Newtonian picture of the universe. During the period from 1933 to the outbreak of the war they had some success in persuading Nazi Party chiefs of the greater conformity of German Physics with Nazism, but they had few converts: by the end of 1939, adherents of German Physics held only six of the eighty-one professorships in physics available in Germany or Austria, with their numbers declining thereafter. In 1939, the movement did succeed in denying Werner Heisenberg, the prominent quantum theorist, the chair at Munich in favour of a German Physics follower, Wilhelm Muller, but the victory was short-lived. As a demonstration of the power of Party connections within the polycratic character of the German state, Heisenberg was rehabilitated by the use of family links with Heinrich Himmler. This served to demonstrate the fact that opposition to German Physics was not seen as opposition to the regime.[156] The Nazi Party's faint interest in internal disputes within the sciences was illustrated by its largely ignoring an attempt to found a Deutsche Mathematik movement.[157] Surprisingly, a movement for German biology did not develop, the absence of which Ute Deichmann attributes to the state of biology at the time.[158]

Ironically, it was to be Heisenberg and other victims of the German Physics movement who were to be among the leaders of the forlorn German attempt to develop an atomic bomb. This reversal points to the fact that German Physics was largely discarded come the war, when the rejection of modern physics was seen as a potential obstacle to war-related research.[159] The availability of physicists of the calibre of Heisenberg also shows that, despite the purge of Jewish scientists, and particularly physicists, the field still had major figures active in Germany. The very basis of the quest for the uranium bomb was the demonstration in 1938 by the German chemists, Otto Hahn and Fritz Strassmann, that commonly occurring uranium-235 could be split under radioactive bombardment – work that was performed at the KW Society Institute for Chemistry in Berlin. Most scientists retained the apolitical stance they had adopted during the Weimar regime, distancing themselves from the nature of the state under which they worked. As with so many regimes, scientists proved capable of doing good work so long as they were not actively persecuted and given sufficient funds to conduct their work.[160]

[156] Josephson, 'Science, ideology, and the state physics', p. 589
[157] Walker, *Nazi science*, p. 87   [158] Deichmann, *Biologists under Hitler*, p. 82
[159] Walker, *National socialism*, pp. 81, 85   [160] Walker, *Nazi science*, p. 271

Apart from racial theorising, science was in general largely irrelevant to the Nazis in ideological terms. Science was chiefly valued for its practical advantages. Few German scientists were Party members, and Nazi ideology valued the valour of soldiers over the meticulous work of scientists. Scientists were generally from the privileged classes, in contrast to the Nazi officials. The Nazi leadership, however, recognised that science and scientists were a significant part of Germany's heritage, and so merited support – particularly if there was the possibility of practical advantage. Hence, projects with some promise of supporting autarky or rearmament were favoured; so, too, was work on racial hygiene to reinforce the governing ideology.[161] Under the Nazis, mechanisms were put in place to provide grants for scientific research. The Weimar Emergency Foundation became the German Research Foundation, and in 1937 a Reich Research Council was established. This Council had no particularly strong ideological character and resembled that of countries as diverse as the USSR and the United States. Nonetheless, the attempt by the Council to centralise funding was undermined by the polycratic character of the Nazi state, with different groups using different power centres to advance their interests.[162] As had been true under Weimar, the reluctance of industrialists to conduct their own research led to more state investment in the KW Society, which became a willing servant of Nazi goals. This state largesse continued during the war, in which the Society was a major contributor to the war effort, as it had been in World War I.[163] The lessons of World War I were a continuous reminder to the Nazis of the need to invest in scientific and technological research. The result was a sharp increase in the amounts devoted to research and development from the Nazi takeover.[164]In many instances, such research was carried out through contracts with scientific bodies such as the KW Society.

The Nazi state was by its nature centralising. It denied, for example, many of the powers traditionally enjoyed by the German Länder or constituent states focussing ultimate decision-making on the power of the Führer and his central authority.[165] Consistent with this, the German state was insistent that all German institutions should be under its sway, and this included scientific institutions. The degree of control, however, could differ with different institutions. Universities being subject to the Civil Service provisions fell readily under Nazi state control. Under the

[161] Heim, Sachse and Walker (eds.) *Kaiser Wilhelm Society*, p. 2
[162] Walker, 'Twentieth century German science', pp. 804–6
[163] Grunden et al., 'Laying the foundation', p. 92
[164] Grunden et al., 'Laying the foundation', p. 105
[165] Pfetsch, 'Germany: three models', pp. 189–208, 196

Nazis, university rectors were state appointees rather than being elected.[166] The Berlin Academy of Sciences and Humanities, which dated from the time of Frederick the Great in the eighteenth century, was an honorary body somewhat peripheral to the actual conduct of science and was left relatively undisturbed for a few years after the Nazis came to power.[167] Thereafter, however, it was comprehensively made subject to the state to the extent that, from 1936, its members were appointed by the state rather than being elected. By 1939, it was effectively part of the German state and was to benefit from the wartime looting of conquered nations.[168]

By contrast, the KW Society was allowed greater autonomy so long as it generally cooperated with the state.[169] The difference reflects the extent to which the KW Society under Max Planck was generally willing to accommodate the proceedings of the Society to Nazi demands while retaining autonomy over other areas. The KW Society also learned to use the Nazi polycratic state to its own advantage, receiving support, for example, from some industrialists.[170] Another consideration for the KW Society was the need to keep its membership as united as possible in the face of its varying political sympathies. Its success in walking the tightrope of some concessions and some autonomy helped to achieve this goal.[171] As a relatively minor institution, the German Physics Society was also granted some degree of autonomy, but this lessened with the outbreak of war. In return, the Society was expected to foster research on armaments rather than curiosity-driven research.[172]

Once war broke out, whatever freedoms scientists had retained were more and more restricted.[173] Huge sums were invested in developing spectacular new weapons, such as long-range rockets, but scientists were also involved in more pedestrian tasks. As in World War I, one of their main preoccupations was producing substitutes for goods not available because of the war, including metals, rubber, mineral oil and explosives.[174] During World War II, the Germans became world leaders in the development of synthetic fuel.[175] With the belief in an early victory and the resources which the conquest of much of Europe gave Germany, there was less incentive in the early stages of the war to invest in programmatic scientific research. A telling example was the way in which research on radar ceased in 1940 and

---

[166] Deichmann, 'Biologists', p. 11
[167] Nötzoldt and Walther, 'The Prussian Academy of Sciences', p. 441
[168] Walker, *Nazi science*, pp. 88, 111, 113
[169] Mark Walker, 'Twentieth century German science', pp. 802–3
[170] Macrakis, *Surviving the swastika*, p. 72   [171] Beyler, 'Maintaining discipline', p. 254
[172] Dieter Hoffmann, 'Between autonomy and accommodation'
[173] Walker, *National socialism and German physics*, p. 86
[174] Grunden et al., 'Laying the foundation', p. 82
[175] Stranges, 'The US Bureau of Mines' synthetic fuel', p. 59

was only renewed in 1942. This was to change after the Russian campaigns and as the war turned against the Nazis.[176] Reflecting this, Albert Speer, Hitler's Reich Minister of Armaments and War Production, began to centralise scientific activity more – the general tendency on the part of a state when the needs of war are paramount. This was, however, in contrast to the German tradition and hitherto Nazi practice of allowing several rival scientific institutions to compete for finance.[177] Growing centralisation also reflected the growing scale of the war projects, notably the ballistic rocket program. Like the Manhattan Project in the United States, the quest for such 'wonder weapons' was a form of 'Big Science' only sustainable with the support of the state.[178]

An instance of the trend towards growing centralisation in the latter part of the war was Hitler's decision in 1942 to create a Reich Research Council, which would be supported directly by the state and operate under the supervision of Hermann Göring. Thus, scientific matters became part of the concerns of the workings of the Party. The Nazi state also initiated a greater degree of central planning in the use of its human scientific resources: at the end of 1943, five thousand scientists were brought back from combat roles. By the end of the war some 15,000 German soldiers had been similarly restored to their civilian scientific roles. The need to adopt such measures was in part the realisation that Germany's enemies were effectively using science to wage war, prompting Germany to do likewise. Indeed, Albert Speer, Hitler's Minister of Armaments and War Production, was particularly generous with grants to the KW Society because of fears that German science was falling behind that of its rivals.[179] Such fears may have been voiced because of the extent to which, under the Nazis, science was expected to be applied science, overshadowing fundamental theoretical concerns.

While the Nazi state had a varying impact on the conduct of scientific research, its intrusion into the lives of the population as a whole was considerable and often deadly. Racial theory brought with it the Holocaust, with antisemitism being the most fundamental tenet of Nazi ideology. In this warped world view, racial theory was, as Robert Proctor argues, conceived of in medical terms, with concentration of Jews in ghettos, for example, seen as a form of quarantine. The need to police the social hygiene of the state extended to other groups, such as homosexuals or gypsies. The same mentality of protecting the population against hazards was also seen in a less malign form in the Nazis'

[176] Phinney Baxter, *Scientists against time*, p. 7
[177] Macrakis, *Surviving the Swastika*, p. 96
[178] Neufeld, 'The guided missile', p. 51; Albrecht, 'Military technology', p. 97
[179] Macrakis, *Surviving the swastika*, pp. 91, 94, 154

preoccupation with the impact of environmental hazards, including the effects of smoking.[180]

Notions of 'racial hygiene' were not unique to the Nazis and had been common within the German medical profession before the Nazis came to power. One goal shared by many doctors in the Weimar regime was the implementation of eugenic policies in the interests of greater 'racial purity'. This movement towards eugenics as a form of social planning meshed with the steps taken to expand the welfare state under Weimar – particularly its emphasis on public health.[181] The ground, then, was well-sown for the actual implementation of eugenicist policies when the Nazis came to power. In justification, the Nazis could invoke racial theorising together with the support of a large proportion of the medical profession, thus giving a scientific gloss to their policies. From the time they came to power, the Nazis made plain that it was the state that determined policies concerning marriage and birth, since these in turn determined the nature of the race. The Sterilisation Act of 1935 was passed along with other laws banning cross-racial marriage or sexual relationships. The aim of the Sterilisation Act was to minimise 'worthless lives' by ensuring that those considered deficient could not reproduce. Doctors were obliged to report any of their patients who fell into such a category, obliging them to go before the sterilisation courts established by the law. So extensive were the workings of this regime that by 1945 half a million people had been sterilised. In a move that foreshadowed the workings of the 'death camps', in 1939 the Nazi regime took things to another level by initiating a policy of 'euthanasia' for those living 'worthless lives'. By 1945, 200,000 people died under the provisions of this policy.[182]

Though the use of 'euthanasia' was, fortunately, unique to the Nazis, the import of their eugenics policy and the use of sterilisation (albeit on a lesser scale) had parallels elsewhere. As we have seen, the United States was a pioneering force in legislating for sterilisation, and its laws had an effect on the Nazi legislation.[183] The US and German cases shared some of the same justification in using supposedly scientific criteria to produce a more orderly and better planned society. Science had become such an authority system that it could be used for such ends even when the connection of science with such belief systems as 'racial science' was illusory. To add to the standing of 'racial science', however, the Nazis supported research in related areas, such as genetic and comparative physical anthropology – support which gave rise to over a dozen new

---

[180] Proctor, *Racial hygiene*, pp. 7–8
[181] Porter, *Health, civilization, and the state*, p. 200; Weindling, 'Medicine and modernisation', p. 285
[182] Porter, *Health, civilization, and the state*, p. 193   [183] Proctor, *Racial hygiene*, p. 7

medical journals and a range of scientific institutes.[184] Another form of authentication of Nazi goals was the involvement of prestigious research institutes, such as those of the KW Society, in the euthanasia program.[185]

Under the Nazi state, science was to play a number of roles. The acquiescence of the vast majority of scientists in working for the state gave an element of legitimacy to the Nazi regime. Many such scientists might draw a distinction between working for the government and working for the state, but this did not change the appearance of support. Science was valued as an expression of the German genius and, most tellingly, for its practical benefits. It had limited value as a support for Nazi ideology, which helps account for the way in which at least some German scientists and their institutes were granted a degree of autonomy – as was most notably the case with the KW Society. The advantages of scientific research were rather lost sight of during the early years of the war with their easy victories, but came into focus again in the period 1942–5. In this latter period, however, the Nazi desire for 'wonder weapons' lead to a vast proportion of resources being devoted to rockets when the military benefits were limited. Nazi ideology and a rational scientific policy could be in conflict. Medicine loomed large in the Nazi ideology, with analogies between the maintenance of public health and the elimination of purportedly diseased parts of society. But, as this analogy suggests, Nazi medicine was authoritarian medicine,[186] with its primary obligation being to the race rather than the individual patient. The Nazi totalitarian state and science could generally work together even if the partnership was often a one-sided one.

Russia

1914–1917

The brief period between the outbreak of World War I and the October Revolution of 1917 was a transforming one for the relations between science and the Russian state. The imperatives of war forced the Tsarist government into imposing greater order and system on the nation's resources, including its scientific resources. The need to do so was reinforced by the shortage of supplies of scientifically produced goods such as chemicals and precision instruments, which had largely come from Germany.[187] The planning necessary to overcome such shortages was

---

[184] Proctor, *Racial hygiene*, p. 283
[185] Heim, Sachse and Walker (eds.), *Kaiser Wilhelm Society*, p. 5
[186] Proctor, *Racial hygiene*, p. 288
[187] Siegelbaum, *The politics of industrial mobilization*, pp. 103–4

to be inherited by the Bolshevik regime when it staged its successful revolution. There was, then, a degree of continuity between the old regime and the new.[188] The outbreak of war prompted the Tsarist regime to turn to the Imperial Academy of Sciences for advice on how best to mobilise resources to conduct the war. Traditionally a body largely concerned with pure research in the humanities and the sciences, the Academy took up the task of reordering government instrumentalities the better to serve the war effort. Its major instrument for doing so was the Commission for the Study of Natural Productive Forces of Russia, which provided a platform for applied research – something which was to be taken further by the Bolsheviks, who used the Commission to continue the long Russian tradition of mapping and exploiting the resources of the vast territories of the USSR.[189] Another development in this late Tsarist period that was to be built on by the Bolsheviks was the use of specialist research units drawing on the example of the KW Society.[190] On the other hand, some of the less centralised committees set up to advance the war effort, such as the locally-based War-Industries Committees that actually challenged state-wide bureaucratic structures, did not survive the October 1917 revolution.[191]

1917–1945

The greatest continuity of all between the Tsarist and the Bolshevik regimes was the continued centrality of the Academy. Founded by Peter the Great as a force for modernisation, it had become tradition-bound and preoccupied with its honorific position. Its activities during World War I had done something to reverse such an image. However, its ranks were largely made up of those who were willing to contemplate the moderate constitutional changes of the February 1917 revolution but not the root and branch reforms initiated by the Bolshevik revolution of October 1917.[192] Yet, despite moves to abolish it,[193] the Academy was to be the Bolsheviks' main instrument for ordering Soviet science and for continuing the centralised control that had been a feature of the Tsarist regimes. Remoulding the Academy to suit its purposes was then to be a feature of Soviet policy in regard to the nation's scientific infrastructure.

Over the course of the 1920s, the Academy adapted to the new regime in some respects with, for example, less emphasis on the humanities

---

[188] Brooks, 'Munitions, the military, and chemistry'
[189] Graham, *The Soviet Academy of Sciences*, pp. 165–6
[190] Kojevnikov, 'The great war, the Russian civil war', pp. 248, 252–6
[191] Lehmann, Morselli, *Science and technology*, p. 22
[192] Vucinich, *Empire of knowledge*, p. 93   [193] Vucinich, *Empire of knowledge*, p. 133

rather than the sciences. But plans for making the Academy truly Bolshevik in character came to a head in 1929. This was the year of the 'Great Break': Stalin's ushering in of a new economic plan and, with it, the acceleration of industrial and agricultural modernisation. This brought with it renewed pressure to purge the Academy of those not in sympathy with such Soviet planning. Of the academicians elected between 1929 and 1931, one-third were Marxists, and it was these who gravitated to the top administrative posts.[194] The Marxist hegemony over the Academy brought with it, too, a purge of members, with 128 imprisoned and another 520 dismissed.[195]

The Bolsheviks' preoccupation with refashioning the Academy to their own purposes is testimony to how central was science to their conception of a properly-ordered state. Nazi Germany may not have closely linked science with its governing ideology, but the Soviet state emphatically did. The dominant ideology that, under Lenin's guidance, drew together Marxism and dialectical materialism accorded science a particularly exalted role as the one true path to truth. Forms of science which were seen as weakening the materialist basis of Marxism-Leninism could, consequently, be censored. Science lay at the heart of the national ideology and at the regime's hopes of true progress through rational planning. There was, then, a strong emphasis on the practical applications of science, a source of continuing conflict between the Soviet government and the Academy. Promulgating the national ideology of science even took the form of providing state assistance to provincial science societies as a counterweight to pre-Bolshevik tradition.[196]

The Soviet Union, then, was unparalleled in the emphasis it placed on science as both a foundation of the ideology on which the state rested and as a source of guidance for practical action. The extent of this commitment was underlined by the resources which were poured into science at a time when resources were particularly limited because of the disruption caused by the October Revolution.[197] During the 1920s and 1930s, more support was devoted to science by the Soviet Union than by any other government.[198] While in the 1930s the USSR spent 0.6 per cent of national income on research and development, the United States spent only a little over half of this with a figure of 0.35 per cent.[199] Consistent with the ambitions the Soviet state held for science, early in its history it established specialist research institutes that would complement the work

---

[194] Vucinich, *Empire of knowledge*, p. 129
[195] Graham, *The Soviet Academy of Sciences*, p. 121
[196] Andrews, *Science for the masses*   [197] Graham, *Development of science policy*, p. 13
[198] Graham, *Science in Russia*, p. 17
[199] Lewis, *Science and industrialisation in the USSR*, p. 17

of the Academy. These drew on such foreign models as the German KW Society, but also had indigenous characteristics of their own with close connections with particular industries. By the mid-1930s these institutes were, along with the Academy and educational institutions such as the universities, part of the established landscape of Soviet scientific infrastructure.[200] The Academy served to draw these different sectors together. Whereas in Tsarist times it had dealt almost exclusively with theoretical issues, under the Soviets it also became involved in technical problems and was a major source of research and development for the country as a whole.[201] During the 1930s, too, its share of graduate degrees began to increase.[202]

By the 1930s, the imprint of Stalin was increasingly evident in the workings of Soviet science. Stalinist science was conceived on a national scale, requiring huge resources and the cooperation of scientists across the country. Driving the whole system of science under Stalin was the most fundamental feature of communist rule: the use of planning coordinated by state agencies.[203] The Soviet scientific establishment had grown from a modest position under the Tsars to one of the world's major scientific centres. Given that the state was the only possible patron, science was, from the viewpoint of government, 'Big Science' with the individual scientist subordinate to a large complex of institutions.[204] The increasing investment by the state meant that more and more obstacles were put in the way of scientists maintaining their international connections. While in the 1920s such connections still were permitted, in the 1930s Russia began to close its scientific borders – an important manifestation of this being the Academy's decision not to elect foreign fellows after 1935. Another indication of the sundering of international ties was the officially decreed break with the Rockefeller Foundation in 1933.[205] Russian science, as Nickolai Krementsov argues, was increasingly seen as distinctively national with a proletarian and materialistic character.[206] Together with such a nationalist view of science there was also the characteristic Stalinist fear of political contamination if scientists had too much contact with the West. From their very different premises, both totalitarian ideologies – communism and Nazism – both came to oppose the internationalism which was basic to the normal workings of science.

[200] Graham, 'Development of science policy', p. 29
[201] Lewis, *Science and industrialisation*, pp. 26–7
[202] Vucinich, *Empire of knowledge*, p. 149   [203] Krementsov, *Stalinist science*, p. 42
[204] Krementsov, *Stalinist science*, p. 3
[205] Solomon and Krementsov, 'Giving and taking'
[206] Krementsov, *Stalinist science*, p. 44

The increasing nationalisation of science brought with it greater centralised control through the Academy, which was integral to the state's relations with the scientific community. Symbolically, in 1934 the Academy was moved from Leningrad (formerly St Petersburg) to the post-revolutionary Soviet capital, Moscow, where it was close to the state's central institutions. Another sign of increasing state recognition of the Academy as the capstone of the scientific establishment was that in 1936 it absorbed its rival, the Communist Academy.[207] This was founded in 1918 as the Socialist Academy and then continued from 1924, when it was renamed the Communist Academy. The Communist Academy, as its name suggests, had been maintained by the Soviet state as a means of promoting Marxist ideology particularly through study of the social sciences and humanities, but it also came to encompass the study of the natural sciences. Having two rival institutions for a time suited the Soviet state, especially as the Communist Academy was ideologically more punctilious. For Stalin, however, having that degree of autonomy in a centralised system was probably considered a possible danger to be eradicated by making the Academy of Sciences paramount. Stalin also eliminated professional associations of scientists as another possible source of division.[208]

The Academy came, then, to embody par excellence the distinctive features of the Soviet scientific system in integrating the conduct of science with the mechanisms of state control. It was a tribute to the malleability of the Academy that it had not only survived but had vastly expanded under Soviet rule. As Loren Graham points out, it was the only one of the scientific academies created in the eighteenth century that still largely reigned over its national research in the twentieth century.[209] The larger and more complex the system, the more possibility there was for individuals to manoeuvre between the cracks so that, though certainly totalitarian, the Soviet system was not monolithic.[210] Stalin regarded science in largely practical terms, echoing the Bolshevik slogan 'Science in the service of socialist construction'.[211] He also sought to promote what he called proletarian science that arose from below rather than from the traditional hierarchies.[212] Enforcement of his views came to a head in the period of the Great Terror from 1936. The purges were used to eliminate from the Academy those too wedded to 'bourgeois' notions of science, as well as those considered guilty of ideological unorthodoxy.

---

[207] Krementsov, *Stalinist science*, p. 37   [208] Josephson, 'The political economy', p. 147
[209] Graham, *The Soviet Academy of Sciences*, p. 191
[210] Krementsov, 'Russian science', p. 777
[211] Krementsov, 'Russian science', p. 782; Johnson, *Stalinist politics*, p. 260
[212] Josephson, *Physics and politics*, pp. 141, 184

Stalin's dominance over the Academy was made manifest by his election as an honorary member of the Academy in 1939.

With Stalin's dominance came an increasing insistence on ideological conformity. This reflected the importance science held for Stalin as a way of confirming Marxism-Leninism.[213] Ironically, persecution of scientists reflected the importance that Stalin placed on science. Two main areas of attack on modern science stood out: firstly, forms of post-Newtonian physics such as quantum mechanics, and secondly, Mendelian genetics. Modern physics in the eyes of its Stalinist critics appeared to undermine a consistently materialist explanation for the workings of nature. By contrast, Newtonian mechanics with its basic framework of particles in motion was much more consistent with Marx's view of reality. Use by modern physicists of statistical and probabilistic descriptions was seen as a form of 'idealist' departure from the bedrock materialism of Marxism; so too was the substitution of mathematical concepts for physical explanations. Ironically, some of the objections to modern physics mirrored the critiques mounted in the name of the defence of German Physics by Lenard and Stark in Nazi Germany. There was even an anti-Semitic undertone to the Soviet as well as the German assult, with Soviet Jewish physicists being singled out for attack. Soviet relativist physicists argued that it was impossible to maintain the Newtonian world view in the face of the mounting experimental evidence against it.[214] It was pointed out, too, that the Soviet critique paralleled that of the ultra-Nazi German Physics movement.[215] The most compelling repost of the relativists came with the need to undertake research for the atomic bomb during World War II, which was impossible to do without invoking recent modern physics. This quietened the critique for a time, though the attack on 'idealism' in physics was to break out again after the war.[216]

The attacks on modern genetics attracted more widespread attention both within the Soviet Union and abroad. This conflict was fought with particular intensity for a number of reasons: first, Stalin claimed to have some competence in the science of plant breeding, which was the battle field on which the attack on Mendelian genetics was fought. Secondly, the views of the main anti-Mendelian, Trofim Lysenko, developed in the early 1930s, about the adjustment of plants to their environment and the passing on of characteristics acquired from this interaction with the environment, appeared as an example of proletarian science. Such a view was heightened by the fact that Lysenko was the son of peasants and on

---

[213] Pollock, *Stalin*, p. 3   [214] Josephson, *Physics and politics*, pp. 246, 273
[215] Josephson, 'Science, ideology, and the state physics', p. 585
[216] Josephson, 'The political economy', p. 152

the periphery of the scientific establishment. Stalin, too, was responding to the campaign for collective farming and the hopes of increasing yields that Lysenko seemed to encourage.[217] Though Lysenko did not write about human evolution, his anti-genetic approach seemed more consistent with the Marxist commitment to progress than the bleak view that many of the characteristics of plants and animals were determined before birth.[218] Indeed, one form of attack on the orthodox Soviet geneticists was that their views resembled Nazi racial determinism.[219] Such an outlook helps to explain why eugenics, after some early interest as a form of social planning, did not gain ground in the USSR, being too closely identified with Nazi ideology.[220] Among the strongest Soviet critics of eugenics was Lysenko.

Stalin, then, was deeply committed to the defence of Lysenko's ideas as a distinctive form of Soviet science which meshed well with the theory and practice of the Stalinist state. It was also a statement by Stalin of his right, as custodian of the state ideology, to determine what was valid science.[221] The high point of Stalin's support for Lysenkoism came in 1948 with its official approval, making it clear that the state had the right to determine what was 'true science' as part of its larger duty to determine the true philosophical principles on which the regime was based.[222] Lysenkoism, then, was part of a state-defined science that was intended to provide the foundation for the state's needs with an emphasis on practice rather than theory.[223] The result was relentless persecution and even execution of Lysenko's opponents. The campaign was to last until Stalin's death in 1953 and to continue on into that of Khruschchev's regime. With Khruschchev's dismissal in 1964, Lysenko came under increasing scientific attack and his influence faded. So linked was Lysenkoism with the defence of orthodox Marxism-Leninism that it was exported to Communist China, with disastrous results when its principles were put into practice with deadly literalism.

The purges of the late 1930s proclaimed the authority of the Party and its leader to proclaim what was acceptable science. Yet, despite the purges, Soviet science remained in a sufficiently strong state to respond to the war with resilience and purpose. In some ways the Soviet Union was better fitted scientifically for war than some other regimes. Its scientific establishment

---

[217] Graham, *Science in Russia*, p. 126
[218] Josephson, *Totalitarian science and technology*, p. 24
[219] Krementsov, *Stalinist science*, pp. 73–4
[220] Graham, *Science, philosophy, and human behaviour*, p. 128
[221] Holloway, *Stalin and the bomb*, p. 366
[222] Graham, *Science, philosophy, and human behaviour*, p. 15
[223] Josephson, *Totalialitarian science and technology*, pp. 7, 24

had already gone through the restructuring and centralising measures that other states, such as the United States, had to conduct in the context of the war itself.[224] Indeed, the war led to an even more central role for the Academy as it supervised wartime technical research programmes.[225]

War brought some relief from the ideological pressures to which Soviet scientists had been subject.[226] The need for political and military chiefs to follow the advice of scientists loosened the regime's grip on science.[227] At the same time, the contributions science was making to the war effort strengthened relations between science and the state, one important manifestation of which was government fulfilment of most scientifically-based requests for resources. The Academy reciprocated with a shift from fundamental to military science.[228] Soviet science responded to the war with incredible tenacity, absorbing the changes required for the removal of much of their research plant to the Urals from 1941. The greater wartime autonomy of Soviet science led to some lessening of its international isolation with some scientific interchange with its allies in the fight against Fascism. One indication of such growing internationalism was the decision once again to allow election of foreign members to the Academy in 1942.[229] The needs of war brought with it a great expansion in the size of the Soviet scientific estate,[230] much of which was retained after the war in response to the Cold War.

Despite the devastating German invasion, Russia made some progress on research on developing an atomic bomb during the war.[231] This was an indication of the strength of its nuclear physics community, which in the 1930s was the largest in the world.[232] Such a background explains how Soviet scientists could produce their own atomic bomb four years after Hiroshima. Working on this scale was a natural feature of the Soviet scientific system as it developed in the 1930s. The trend towards 'big being beautiful' was strengthened by the fact that many of the Party officials in the 1930s had had a technical education, which naturally inclined them towards large-scale projects.[233] The end of the war brought with it the need to develop new patterns of cooperation between the Soviet state and the scientific community, especially given the national priority of producing an atomic bomb. That project was put under the control of Lavrentiy Beria, head of the secret police, who evidently

[224] Graham, *Development of science policy*, p. 30
[225] Lewis, *Science and industrialisation*, p. 78
[226] Vucinich, *Empire of knowledge*, p. 198   [227] Krementsov, *Stalinist science*, p. 96
[228] Krementsov, 'Russian science', p. 787   [229] Krementsov, *Stalinist science*, p. 115
[230] Krementsov, 'Russian science', p. 786
[231] Holloway, *Stalin and the bomb*, p. 89; Holloway, 'Entering the nuclear arms race', p. 160
[232] Agar, *Science in the twentieth century*, p. 313
[233] Graham, *Science in Russia*, pp. 164–5

needed to be reminded of the need to win over the scientific community rather than simply order them about. Interestingly, when conveying that warning to Stalin in October 1945, the eminent nuclear physicist, Peter Kapitsa, invoked the authority of Francis Bacon for maintaining the importance of scientists in the body politic: 'Without that patriarchal position for scientists the country cannot grow culturally on its own, just as Bacon noted in his New Atlantis. It is time, therefore, for comrades of Comrade Beria's type to begin to learn respect for scientists.'[234] Such warnings seem to have had an effect as the project was successfully completed through the collaboration of the central government in the formidable person of Beria. Recognition of the nuclear scientists working on the project took the highly tangible form of being granted much-valued privileges when it came to housing and access to food rations.

The success of Soviet science in adapting itself to war was the product of developments that had dated back as far as World War I. In the expiring years of the Tsarist regime there was some attempt to muster the nation's scientific resources more effectively. This was a small prelude to the fundamental changes that were to occur in the position of science in the Soviet post-1917 reconstruction of the nation. Science now was an integral part of the Marxist ideology that legitimated the regime. It was also central to the great faith in the possibilities of development and, with it, rational planning. Planning brought with it an ever increasing role for the state so that almost no activity, including science, was without its political element.[235] The very importance accorded to science meant that it was vulnerable to attack, particularly as ideological debates reached fever pitch in the late 1930s. Yet, for all the damage done to science, the Soviet scientific community continued on with its endeavours, giving some pause to the assumption that science needs freedom to operate[236] (though Stalin did set back Soviet genetics). Though the outbreak of World War II followed on the heels of the Great Terror, the German invasion found the Soviet Union with an extensive and well-organised scientific establishment that ably supported the victorious war effort. Furthermore, it was able to produce a Soviet atomic bomb by 1949. Soviet science was to demonstrate the resilience of science, even though its institutional forms were later to prove an obstacle to scientific productivity. Science had become too large, too well-organised and too essential to the civilian and military life of the state to be badly damaged even by a state with which it was meant to be in alliance.[237]

---

[234] Agar, *Science in the twentieth century*, p. 315
[235] Beyler, Kojevnikov and Wang, 'Purges in comparative perspective', p. 31
[236] Loren R. Graham, *What have we learned*, p. 131
[237] Loren R. Graham, *Science in Russia*, p. 198

Conclusion

By its nature, war accentuated the power of states to command and to reorder, and this was to change their relationship with science as with much else. Different states adapted in different ways. The United States, with its tradition of decentralised, federal government, made only a partial adjustment to the needs of war in World War I. What scientific innovation there was, was closely tied to the armed forces. A National Research Council was created as a wing of the Academy of the Sciences, but was not given state support. But, then, the United States only belatedly became a combatant in World War I and its national interests were not as closely engaged as in World War II. By the 1930s the USA had become a major world scientific power, but a lot of the credit for this belonged to the foundations (especially the Carnegie and Rockefeller) rather than the federal government. World War II meant, then, a very rapid construction of institutions to supervise federal government expenditure on scientific research. Of these, the most important was the Office for Scientific Research and Development, which balanced traditional suspicion of centralisation and the power of the state with the needs of war through the use of the scientific contract. Thus, scientists generally continued to operate in their familiar environment rather than the research being done in giant in-house state institutes. Wartime science, then, brought about a considerable expansion of the US scientific estate, but in a form which was acceptable to US sensitivities about the role of the state.

As a fellow liberal democratic state Britain faced some of the same issues, though the need to reach a resolution was greatly accelerated by its much more whole-hearted involvement in World War I. Its solution was a government agency to coordinate the conduct of science throughout the country. Though discussed frequently with various proposals for an overarching Department of Science, this was not an approach ever adopted by the United States. When founded in 1916, however, Britain's Department of Scientific and Industrial Research attempted to ensure that the direct power of the state was buffered by in-between committees, with the goal of separating supervision of research from administrative oversight. The DSIR was sufficiently successful in combining responsiveness to Britain's wartime needs with sensitivity to British suspicion of too much state power to be maintained in peacetime. Come World War II, then, Britain largely had its scientific infrastructure in place, in contrast to the United States. This helped facilitate the active involvement of the scientific estate without too much friction about the expansion of the wartime state's powers.

During World War I, the French, like the British, had to create in the conditions of wartime a functioning research infrastructure. Their main instrument for doing so was the Directorate of Inventions, which reflected the nature of the state in centralising power to promote efficiency in wartime planning. Balancing the expansion of the state-oriented science with reservations about the expansion of state control was, then, less of an issue in France than in the USA or Britain. The success of the wartime experience in directing science for military purposes led between the wars to a variety of bureaucratic experiments for the funding of science and research generally culminating in 1939 in the Centre National de la Recherche Scientifique (CNRS). Overshadowing such experiments was the imperative to organise France for another war, which duly came but in a way that left little room for France to deploy its new structures for linking science with the state.

One of Germany's great advantages in World War I was that it had an effective scientific infrastructure in place, particularly with the Imperial Physical and Technical Institute and the Kaiser Wilhelm Society. War led to an expansion of such instrumentalities and a redefining of some of their purposes, but did not require a wholescale revision of the organisation of science in a way that was true in France and Britain. War, however, did mean more scientific dependence on the state. This continued to be true in the Weimar period, when Germany's troubled circumstances created the need for expanded state funding of the sciences. Nazi rule brought with it further state control over science and more direction about the ends for which scientific research was conducted. In a state where the claims of various power blocks had to be balanced, however, there was the possibility, as the KW Society showed, of retaining some degree of autonomy in return for cooperating with the state's decrees. War expanded the state's control over the conduct of science, but the full impact of this was delayed to around 1943 after Germany's rapid victories began to be reversed. Nazi ideology demanded war, and science as an integral part of Germany society was expected to play its part – which most scientists did without demur.

For the other major totalitarian power, the Soviet Union, science was much more an essential component of its regime. Science was seen as an embodiment of the governing ideology of dialectical materialism. Furthermore, science provided the means of producing the material progress that the state's Marxist ideology promised. Such a view of science led to the comprehensive reshaping of the Tsarist scientific establishment, and particularly the Academy of Sciences. It also required from the state an immense investment to bring science rapidly up to world standards. The support of science, then, loomed larger as a commitment by the state than in Germany or, indeed, in other nations at the time. Expansion of science was accompanied by the nationalist view that Soviet science had a particular character that set it apart from that of the West. Similar views and a shared suspicion of political contamination led both totalitarian powers to cut their scientists off from international contact. The often brutal reordering of the Soviet scientific establishment between the wars meant that when World War II broke out the state was able effectively to command its scientific resources.

The Soviet Union was the country where the alliance of state and science went furthest, with no other patron for science than the state. The Bolshevik revolution had made the state the central organ of the society as a whole, and this was reflected in the way in which science was organised, with a dominant Academy acting almost as a state agency. War exacerbated this trend but it was well in place before World War II broke out. In Germany, state control over science had also largely been established before the war began so that again there was no fundamental change in the scientific infrastructure during the war. In countries with a tradition of liberal democracy, like the United States and Britain, the extent of the state's control over science was a much more qualified one. Britain also had much of its scientific infrastructure in place when war broke out, but this included provisions to maintain as far as possible a degree of autonomy in regards to scientific research. For the United States, with a much less cathartic involvement in World War I, the adjustments to its research structure had to be made rapidly in the midst of war. This necessitated some measure of centralisation, but the whole *modus operandi* of US wartime research was to attempt to offset this with as much of its research as possible being farmed out to large firms and universities.

Whatever their response to the problem of effectively linking science and the state, these different countries all reflected the need to redefine the role of science in a polity in which war had vastly expanded the power of the state. Science also had expanded its effectiveness and utility as a weapon of war, making states determined to incorporate it into their wartime machinery. There were few scientists that resisted the process, underlining the extent to which science can operate in very different political settings. Such was the strength of the symbiotic partnership between the wartime state and science that it survived into peacetime. This was truer after World War II than World War I. In both, science played a key role, but World War II was fought with weapons that reflected recent innovative research, whether it be rockets, radar or, above all, the atomic bomb. World War II, then, brought the symbiotic relationship between science and the state to a level that was to make the partnership a permanent one. Francis Bacon's dream of a state wedded to scientific progress had been realised, but at a wartime cost that would have been beyond his comprehension.

8   Science, the State and Globalisation

Introduction

The development of the modern state was accompanied by a complementary process: the emergence of a new international order. States do not exist as singular entities; rather, they define themselves in terms of other states and seek to gain a place in the larger international order by their relations, good or bad, with other states.[1] Traditionally, the Treaty of Westphalia (1648), at the end of the Thirty Years' War, is regarded as a watershed when Europe acknowledged that any stable ordering of national claims had to accept that this was a matter of arbitrating between sovereign nations[2] rather than invoking the supernational power of pope or emperor. Such a view of the world order was transmitted by the West by imperialism and trade to much of the rest of the world. Formation of a state became part of the price of admission to the international order.[3] Ironically, however, states could also be agents of anticolonialism[4] providing, as they did, the mechanism to mobilise the peoples and resources of a particular region against the imperial power. In any case, a state, with its claims to sovereignty, could not readily coexist with an imperial regime that demanded ultimate authority.

Closely allied with the spread of state structures and an international order based on states was an alliance between states and science. To establish themselves, states needed the order and system that science could provide, especially in the form of the statistical sciences with their devices for measuring the population and its needs. The very definition of a state and its territories owed much to scientific mapmaking,[5] and the tallying of a nation's assets to the investigations of geologists, botanists and zoologists. Commanding the loyalty of their people meant that the

[1] Vincent, *Theories of the state*, p. 225
[2] Nelson, *The making of the modern state*, pp. 60–1
[3] Tilley, 'Western state-making', p. 637   [4] Tinkler, 'The national state', p. 119
[5] Moore, *Disrupting science*, p. 12

state had to provide forms of welfare that science, and particularly medicine, could provide. In an international order that had grown out of the
need to deal with perennial conflicts between states, both new and old
states sought to arm themselves with the deadliest weapons that military
technology could produce. Maintaining an existence in such stormy
international waters helped shape the character of both the existing and
newly-minted states.

Latin America

Imperialism was an indirect agent of state-making across the globe as
empires eventually collapsed and produced states in their place.
The earliest European empires, those of Spain and Portugal, disinte-
grated in Latin America in the 1809–26 Wars of Independence, with
the metropolitan powers weakened by the Napoleonic Wars. Among
those active in leading the revolt against Spain were graduates from
some of the scientific training institutes set up by Bourbon Spain – such
education made creoles particularly conscious of their subject status.[6]
As was true in other empires, imperially-sponsored education could be
a subverter of empires. In place of the vast Spanish territories under
viceroys arose a cluster of states and, in place of the Portuguese colony
of Brazil, there emerged in 1822 an independent Brazil under an emperor.
In these territories the hold of science was relatively weak, though the
Spanish territories particularly had benefited from the Bourbon reforms
of the late eighteenth century. In the flurry of the wars of independence,
however, some of the new scientific institutions erected as part of these
reforms fell into disuse: such was the case, for example, with Mexico's
Botanical Garden (1788) and School of Mines (1792).[7] The new nation
states of Latin America were slow, then, to expand their attention to
science.

One of the first to do so was Argentina. Populated chiefly by European
settlers, from about 1860 it imported European scientists to attempt to
prevent Argentina falling too far behind European science. One of the
tasks of these scientists was to strengthen the claim of the state on new
territories through a scientifically rigorous description of their terrain.
Modernisation of science and of the army proceeded hand in hand,
modelled on European examples. In later years, following an army
coup, there was an attempt in 1948–9 to bring science more fully under
military control, with the meteorological service, for example, directly

---

[6] Cañizares-Esguerra, *Nature, empire and nation*, p. 62
[7] Chambers, 'Period and process' p. 301

answerable to the military. Soon afterwards, Argentina imported more scientists and technical experts.[8]

The need to import European scientists was common in Latin America with its aspiration not to fall behind Europe despite the meagre output of its own scientists. From 1870, President Gabriel Garcia Moren of Ecuador sought to shore up his rule and his reputation for technocratic efficiency by importing a dozen German Jesuits. These brought about the goal of turning the University of Quito into an institute of technology. Chile was also very dependent on foreign scientists, chiefly from Germany and France.[9] Such reliance on overseas experts, however, inhibited the emergence of a locally based scientific community, bringing with it a limit on the scientifically based services that the Latin American state could provide. By the 1930s and 1940s, however, the number of scientists had grown sufficiently for there to be lobbying of various Latin American states for funds for basic research. The limits on this were evident, however, in the need to undertake such projects in conjunction with international bodies. More and more the natural home for scientific research was the universities that have dominated Latin American science policy since World War II. The main clients of university research have been large government agencies,[10] affirming the close links between science and the state.

In the former Portuguese colony of Brazil, Emperor Pedro II (who reigned from 1831 to 1889) used scientific institutions to glorify his newly established regime around the middle of the nineteenth century. With imperial patronage, some natural history museums, an astronomical observatory and botanical gardens were erected. Thereafter, however, the Brazilian state was more inclined to demand practical returns for its investment in science in terms of agricultural or medical advances. In particular, in the late nineteenth century it sought improvements in the coffee bean, which was the lifeblood of its trade. Hence the erection of an agricultural research station at Sao Paolo following a German model.[11] Considerable advance was also made in fostering medical research, with the Oswaldo Cruz Institute at Rio de Janeiro (founded 1898) becoming a major centre for the study of tropical diseases. By the 1930s, the universities were emerging as the central sites of research largely thanks to imported scientists, though state interest in such research was limited until the late 1950s.[12]

---

[8] Ortiz, 'Army and science in Argentina', pp. 154–5, 175
[9] Pyenson and Sheets-Pyenson, *Servants of nature*, pp. 368–71
[10] Vessuri, 'Science in Latin America', pp. 848–9, 855
[11] Dean, Warren, 'The green wave of coffee'
[12] Schwartzman, 'Brazil: scientists', pp. 172–3

The downfall of the Spanish and Portuguese empires in Latin America resulted in a collection of new states to add to the international order. These states inherited only a limited scientific capital from the colonial period, which, together with political instability, helps to account for the paucity of scientists and scientific institutions well into the twentieth century. The need for scientists to reinforce the functioning of the state and to provide services for its population was recognised, but was often realised by the import of scientists from Europe. Increasing international contact has, however, brought home to Latin America's rulers the importance of scientific advance as a component of a modern state and, since World War II, more resources have been devoted to building up a scientific community with roots in the local society.

India

The Spanish and Portuguese empires in Latin America collapsed in a relatively short space of time once revolution broke out. The dissolution of the British Empire was a more gradual process that generally avoided warfare. The Indian and Pakistani states that emerged in 1947 were shaped both by British colonial influences and by indigenous traditions. One feature of the new Indian state was its admiration for science as a modernising force, a view enthusiastically advocated by Jawaharlal Nehru, the first prime minister. How far had such an outlook been in place in the period of the British Raj?

Like states, empires need science to function effectively. Science provides the information that enables stable rule, the information on resources that enables effective exploitation and the supply of such things as basic medical services that help legitimise rule. One of the main agencies of the Indian imperial state was the medical services, which could intrude into everyday life in ways such as replacing traditional midwives with Western-trained ones.[13] To bring about such changes, or to gain necessary data, imperial rulers need information from local communities for matters as diverse as supplying the names and specimens of local flora and fauna[14] or to map out local irrigation services. The use of local information to provide forms of bureaucratic control provided a form of contact between the imperial state and forms of indigenous knowledge. The merging of the local and the imperial also helped to enlist the indigenous elite in the activities of the Raj.[15] The growing range and

---

[13] Arnold, *Science, technology and medicine*, p. 90
[14] Watson and Noltie, 'Career . . . of Dr Francis Buchanan', p. 393
[15] Gilmartin, 'Scientific empire', pp. 1127–8

variety of scientific activity pursued by the Indian imperial state prompts Zaheer Baber to write that 'During the mid nineteenth century British India constituted the site of one of the largest state-sponsored science and technology projects undertaken anywhere.'[16]

Forms of indigenous knowledge, however, were not drawn into the formal structures of education that perpetuated Western science. Nor were the classical developments of earlier Indian sciences, such as mathematics and medicine, linked to this Western curriculum.[17] Indians seeking a scientific education were dependent, then, on imperially founded institutions. The first such which was open to Indians was the Hindu College in Calcutta, founded in 1817 with an enthusiastic and numerous student body. For Indian intellectuals, the forms of Western science offered a path to Indian modernisation, a view later echoed by Nehru. Kapil Raj argues that Western science could also serve sectarian ends, with the Hindu Brahmins adopting the new Western forms of knowledge as a way of differentiating themselves from the Muslim elite.[18] There were, however, only limited opportunities for Indians to acquire such scientific expertise – not until 1879, for example, did Bombay University introduce a Bachelor of Science degree. Those Indians who did acquire such qualifications had limited opportunities to use them. When it came to the personnel for such empire-defining scientific services as the Great Trigonometric Survey, or the collection of flora and fauna, the key positions were in practice reserved for Europeans. In the exception that proves the rule, Pramatha Nath Bose in 1880 broke precedent when he obtained a graded post in the Geological Survey (though even within the graded posts there was a salary differential between indigenous and European scientists).[19] A few years before, in 1876, the foundation of the Indian Association for the Cultivation of Science helped to strengthen solidarity among Indian scientists and, over time, to associate the promotion of science with Indian nationalism.[20]

Imperial Indian science in the nineteenth century was generally directly commissioned by the state for its own purposes, but by the early twentieth century the universities were becoming more centres of research. An exception to this was the Institute of Hygiene and Public Health at Calcutta, founded in 1932 as a result of a survey of medical education by the ubiquitous Rockefeller Foundation.[21] The growing number of researchers prompted the formation of a pan-Indian body, the Indian

[16] Baber, *The science of empire*, p. 248
[17] Arnold, *Science, technology and medicine*, pp. 5, 181
[18] Raj, 'Knowledge, power and modern science', pp. 121–2
[19] Basu, 'The Indian response', pp. 126–8   [20] Krishna, 'The colonial "model"', p. 60
[21] Sinha, *Science, war and imperialism*, p. 43

Science Congress, in 1914, a body modelled on the British Association for the Advancement of Science. The appeal of this to a large body of Indians[22] was an indication of the growing dissemination of scientific literacy. The increasing sophistication of Indian science was underlined when, in 1930, C. V. Raman, professor at Calcutta University and former president of the Indian Science Congress, won the Nobel Prize for Physics, the first non-Western Nobel science laureate. By 1934, the Indian scientific community had grown sufficiently to found the Indian Academy of Sciences along with the National Institute of Sciences in the following year. Together with the Indian Science Congress, these two bodies provided direction for Indian science in the absence of a central agency such as the British DSIR.[23]

Science indeed began to shape some of the responses of Indians seeking independence. In the period between the wars Nehru enthusiastically advocated the use of science as an ally of an independent state,[24] a position he maintained when he came to power. Science and Westernisation indeed became points of division within the independence movement, with Nehru and others rejecting the 'back to the village' policies of Gandhi in favour of scientific progress. Such advocates of the Indianisation of science looked with admiration at the planned scientific research of the Soviets. Those of this mind were numerous enough to found a journal, *Science and Culture*, which promulgated views similar to J. D. Bernal's advocacy of the use of greater planning in British science.[25] One response to the strongly held belief in the alliance between science and progress was to see it, as Gyan Prakash does, as a mark of the success of the British in winning over much of the Indian elite to a belief in the possibilities of improvement based on an alliance between science and the state – a belief which was to survive the departure of the British in 1947.[26]

Though such proposals for a planned scientific sector were not realised, Indian science and technology continued to grow. When World War II broke out, the Indian scientific community was one of the largest outside Europe and North America, and India numbered eighth in the world count of industrialised nations.[27] Though, up to this point, there had been no central planning or oversight of Indian science, the war brought with it greater centralisation by the imperial government. Drawing on the British model of the Department of Scientific and Industrial Research, in 1940 the Board of Scientific and Industrial Research was established,

[22] Arnold, *Science, technology and medicine*, p. 161
[23] MacLeod, 'Scientific advice for British India', p. 381
[24] Kapila, 'The enchantment of science', p. 131
[25] Sinha, *Science, war and imperialism*, pp. 51–2   [26] Prakash, *Another reason*, p. 8
[27] Arnold, *Science, technology and medicine*, p. 205

which metamorphosed into the Council of Scientific and Industrial Research in 1942, a body which continues to the present. Though the level of exchange between this body and its British equivalents was not as open as it was with the equivalent Australian and Canadian institutions, it was, nonetheless, significant and helped lift the standard and standing of Indian science. As British victory over the Japanese seemed more assured, research was less exclusively military and turned to civilian issues of postwar reconstruction.[28] Indeed, World War II did much to solidify science as a part of the Indian state as more and more Indians took part in wartime research.

Wartime scientific institutional infrastructure helped establish a platform on which the newly independent state could build. A year after independence, in 1948 Nehru established a Department of Scientific Research (redesignated the Ministry of National Resources and Scientific Research in 1951). Such institution building, which extended to seeking advice from J. B. Bernal,[29] was a part of Nehru's basic attitude to the alliance between science and the state expressed pithily in his maxim that 'the future belongs to science and to those who make friends with science'.[30] One indication of Nehru's determination to keep science under the direct control of the state was the locating of much scientific activity (including nuclear research) in research units rather than universities.[31] The change in state structure brought other changes to Indian science, including new national and international networks as the traditional British ones faded.[32] Independent India's determination to forge new links within the international community was reflected, too, in its scientific alliances.

Canada, Australia and New Zealand

While India did not gain its independence until 1947, the dominion settler states of the British Empire were, in constitutional terms, independent from the time of the 1931 Statute of Westminster (though Australia did not ratify this until 1942, and New Zealand not until 1947). Nonetheless, their ties to Britain remained close and this was reflected in the conduct of their science policy, as of much else. The value of science as an instrument of improvement was part of the mentality of these settler states and part of the justification for taking possession of indigenous lands. One form of taking possession was to map the

---

[28] Sinha, *Science, war and imperialism*, pp. 134–5
[29] Sinha, *Science, war and imperialism*, p. 202
[30] Arnold, *Science, technology and medicine*, p. 209   [31] Kapur, 'India', p. 213
[32] Kapur, 'India', p. 217

resources of the land, and from the mid-nineteenth century colonial geological surveys were established to fulfil this goal. Particularly prominent was the Geological Survey of Canada, founded in 1842 along the lines of the British and US state survey. Possession of the land also required making it as productive as possible. Agricultural research was also an illustration of the practical benefits of science. Hence, the colonial governments established agricultural research stations such as the Dominion Experimental Farm System, founded in Ottawa in 1886.[33]

It took World War I to prompt these governments into establishing state instrumentalities to oversee the conduct of science in the nation as a whole. As loyal subjects of the British Empire, the model adopted was that of the British Department of Scientific and Industrial Research, founded in 1916. Britain strongly encouraged such an emulation, suggesting that the dominions might also facilitate the links between science and industry and 'co-operate with the Home-land in the solution of problems of common interest'.[34] Canada was quick to obey, and in 1916 it established its Honorary Advisory Council for Scientific and Industrial Research, which soon became the National Research Council. Part of the groundwork for this had been laid by the foundation of the Royal Society of Canada in 1882, which had helped develop a nationwide scientific network and which was an enthusiastic supporter of the Advisory Council.[35] Australia also acted promptly, and in 1916 established its Advisory Council of Science and Industry. After rather uncertain financial fortunes this was re-established in 1920 as the Institute of Science and Industry. It had another name change in 1926 to reflect the agitation for greater involvement by the six states with a governing council, hence it became the Council for Scientific and Industrial Research. New Zealand did not establish an equivalent body until 1926, naming it the New Zealand Department of Scientific and Industrial Research in deference to its British model.[36]

Although these three government agencies had a common model and purpose, they were to illustrate different ways of linking science with the state. While the Canadian National Research Council was particularly focussed on engineering,[37] the Australian Council for Scientific and Industrial Research paid much attention to agriculture – though becoming more mindful of the needs of industry as war approached in the late 1930s. In 1940, this led to the foundation of a National Standards

---

[33] Kenney-Wallace and Mustard, 'From paradox to paradigm', p. 194
[34] Galbreath, *DSIR . . . New Zealand*, p. 14
[35] Levere and Jarrell, *A curious field-book*, pp. 15, 21
[36] Galbreath, *DSIR . . . New Zealand*, p. 9
[37] Kenney-Wallace and Mustard, 'From paradox to paradigm', p. 195

Laboratory under the control of the Council for Scientific and Industrial Research. Both Australia and Canada had councils that removed them to some degree from the direct scrutiny of government; the New Zealand DSIR, by contrast, was a government department.[38] The Australian institution was led by scientists, while the Canadian one was shaped by the British Department of Scientific and Industrial Research's more bureaucratic model.[39]

All three developed their own research laboratories and did much of their own research, in contrast to the US and, to an extent, British model of farming out research. This in part reflects the fact that the university research facilities of these countries were not as developed as in the United States or Britain, and so were less suitable for the contracting out of research. Canada was in more of a position to find research partners than Australia or New Zealand, which helps to account for the fact that the Australian Council for Scientific and Industrial Research (CSIR) centralised national research more than did the Canadian National Research Council. Australia was also more likely to funnel its research through this one organisation than was Canada. It was a model which suited the needs of countries still in a state of development – hence its adoption by India in 1942, South Africa in 1945 and Pakistan in 1953.[40]

In their different ways, the three dominions took up the legacy of the changing relations between science and the state brought about by World War I. Prompted by the British example, war made evident the need to mobilise scientific resources. This was an impetus that continued into peacetime and helped science achieve a higher profile in government deliberations. In some ways, the Canadian and Australian peak scientific institutions were throwbacks to the academy model: centralised, government funded and linked to the state's practical needs. The difference was the British tradition of creating 'in-between' bodies that weakened the direct power of the state. Such structures could be readily adopted because the dominions' state structures were so close to Britain's. As other bodies, notably universities, became more willing and able to take up the work performed by these institutions, their role diminished. Nonetheless, they played a significant part in establishing viable scientific communities by bringing scientific concerns directly in contact with the apparatus of government. Within the international community they operated chiefly through the networks of the British Empire (as did India before independence), though with some more diverse contacts. Canada had frequent contact with US science, and its French community also

[38] Galbreath, *DSIR . . . New Zealand*, p. 153   [39] Schedvin, *History of CSIR*, p. 25
[40] Schedvin, *History of CSIR*, p. 117

exposed it to some European developments: when its Royal Society was founded in 1883, for example, it drew on the example of the Institut de France. The focus on agricultural research by the Australian Council for Scientific and Industrial Research led it to some partnerships with the most outstanding agricultural research establishment in the world, the US Department of Agriculture. Such contacts were to expand further as the links between Britain and the dominions waned after World War II.

Japan

Latin America and former colonies of the British Empire all represent examples of states formed by the dissolving of empire. Independence brought with it the need to define what role science would play in the new state and how this would relate to the larger international scientific community. Japan offers a very different example as a country that had no colonial scientific heritage, and that shaped the relationship between science and the state in accord with the growth of its other major institutions. In contrast, to the states which emerged from the British Empire, too, it was a country with a strong indigenous national identity – though the form of the state was also shaped by its response to the modernisation it underwent following the Meiji restoration of 1868–1912. After centuries of feudal rule, this, like the absolutist rulers of Europe in their assault on feudalism, brought about a more centralised state – even to the point of establishing a standard version of the Japanese language. Centralisation brought with it the need to define what role science would play in a state that was increasingly defining itself in terms of its programme of modernisation.

Until the Meiji Restoration, Japanese contact with Western science was limited. Dutch traders based in the port city of Nagasaki brought with them some scientific texts, but the then feudally-based Japanese state showed little interest (apart from some attention to astronomy).[41] Among the wider population there was also some interest in Western medicine, but, in general, European influences were shunned as a possible source of instability. When, however, European contact appeared inevitable, the emphasis turned to ways in which Japan could adopt aspects of Western learning that would enable it to resist imperial incursion. A major innovation was the foundation of Tokyo University in 1877, under the control of the Ministry of Education. State agencies that helped define the extent of the state and its resources were the Navy's

[41] Burns, *The scientific revolution*, pp. 151–2

Hydrographic Department, founded in 1871, and the Geological Survey in 1872; a meteorological observatory was also established in 1875.

Science was further incorporated into the structures of government with the establishment of laboratories within the Ministries of Communications and Transportation. The national significance of science was made manifest with the formation of the Imperial Academy of Science in 1908. It was not, however, a major institution for overseeing national research in the manner of the Russian academy.[42] Though the universities were to be the main sites of scientific research until after the war, there were also specialist institutes such as the Institute of Infectious Diseases (founded in 1892 with both public and private funding), which was of international standing and underlined the strong Japanese interest in medicine.[43]

Scientific change in Japan, as elsewhere, reflected the larger context of the nation's relations with other states. Fear of Western imperialism had been the setting for the changes brought about by the Meiji Restoration, while Japan's entries into World War I on the side of the Allies prompted moves for greater scientific self-sufficiency, since German scientifically based products were no longer available. Emulating the national standards laboratories of a number of Western countries, in 1917 Japan established its Research Institute for Physics and Chemistry with public and private support. During the war it also set up an Institute for Aeronautics in 1915, which was attached to Tokyo University.[44] Another effect of World War I was to prompt a move away from adopting German models to those from the United States, which had already been the model for Japanese agricultural research. The foundation of the National Research Council in 1920 also distanced Japan from its previous scientific deference to Germany.[45] The renewed prospect of war in the 1930s led to a greater preoccupation with self-sufficiency that was reflected in the foundation of the Japanese Society for the Promotion of Science in 1932.[46] Hideki Yukawa's publication of his paper on mesons in 1935 (for which he was awarded the Nobel Prize in 1949) marked the extent to which Japanese physics had come of age.

World War II followed the familiar pattern of increasing centralisation, with much of the coordination of military research being conducted through the National Research Council and the Research Institute for

---

[42] Bartholomew, *The formation of science*, p. 123
[43] Bartholomew, 'Science in twentieth-century Japan', pp. 839–87, 879–80
[44] Bartholomew, *The formation of science in Japan*, pp. 198, 217
[45] Bartholomew, *The formation of science in Japan*, pp. 123, 254
[46] Bartholomew, 'Science in twentieth-century Japan', pp. 882–4.

Physics and Chemistry.[47] This, however, was not a marked change for Japanese scientific culture, which reflected the strong intervention of the state in most areas of national life to bring about modernisation.[48] Though much of the scientific work was conducted in universities, this did not greatly lessen the extent to which science was dependent on the state since the universities themselves, founded on a German model, were firmly under the control of the state bureaucracy.[49] Science had been a part of the Japanese restructuring of their state to match the challenge of the West. It continued to bear the imprint of its origins as a state-directed import cultivated for the ways in which it could improve Japanese economic and military standing. Japanese science was not, however, isolated: part of the process of assimilation was the location of appropriate overseas models to incorporate into Japanese structures. Germany was at first favoured, but from World War I and, *a fortiori*, after World War II, the USA loomed larger. Though never part of an empire, Japan reflected the dynamics of an increasingly globalised scientific world.

China

Like Japan, China was never a formal part of an overseas empire, though it had much less success than Japan in avoiding foreign intrusion on its territory. Its contact with Western science went back to the Jesuit missionaries of the seventeenth century, whose astronomical skills were absorbed into the Chinese imperial administration, identifying them with the state. The Jesuits and any such foreign visitors were, however, accommodated on Chinese terms and to the extent that they could serve imperial interests.[50] Growing contact with the West prompted a revival of traditional Chinese mathematics,[51] which was reflected in the way in which the early eighteenth-century Chinese Academy of Mathematics published works in both the Chinese and Western traditions. Increasing Western intrusion in the nineteenth century led to a growing emphasis on the extent to which Western science accorded with traditional Chinese learning and, in particular, that of the Confucian tradition. As mathematics became ever more associated with military power, faith in Chinese mathematics waned as it was associated with Chinese military weakness.[52] Defeat in the Sino-Japanese War of 1894–5 had a particularly devastating effect on confidence in traditional Chinese learning. The war

[47] Bartholomew, *The formation of science*, p. 277
[48] Bartholomew, *The formation of science in Japan*, p. 1
[49] Nakayama, *Science, technology and society*, p. 47
[50] Burns, *The scientific revolution*, p. 147   [51] Elman, *A cultural history*, p. 39
[52] Jami, 'Western mathematics in China', p. 86

strengthened the existing movement to learn from Japan, and from 1902 to 1907 over 10,000 Chinese undertook study in Japan.[53]

Science was associated with modernity and the hope that China might be less subject to foreign intrusion. The imperial reformers had looked to science to achieve such ends, as did the republican regime established in 1911 with Sun Yatsen as president. Some radicals linked the revolution that overthrew the imperial government with the need for a revolution in science.[54] Republics, be they the United States, France or independent India, have a natural affinity for science as a modernising force undermining monarchical or imperial claims. One important symbol of the aspirations to modernity by the Chinese Nationalist (i.e., republican) government was the adoption of the metric system,[55] with its break with traditional custom and adoption of a rationally based measuring system. In constructing the new post-imperial regime, Sun Yatsen and his successor, Chiang Kai-shek, attempted to promote industry using state planning based around a National Resources Commission.[56] This commission also oversaw the emergence of a scientific community that had grown with the establishment of a system of research institutes and the Academia Sinica in 1928, together with an expanded university system.[57] The push towards industrialisation and, with it, application of science-based technology became more urgent as war with Japan seemed imminent. Such predictions proved all too accurate, with the Japanese invasion of 1937 undermining the attempts by the Nationalist regime to establish a new state identity in which science would play a role. The revival of science in China and its links with national ideology were to be suspended until the victory of Mao Zedong in the Chinese civil war and the establishment of the People's Republic of China in 1949. This was to carry further the long tradition of attempting to assimilate Western science into Chinese culture, a quest driven (as in Japan) by fears of foreign domination in a predatory international order.

Nationalism and Internationalism in Science

The penetration of science around the globe underlined the extent to which science can be regarded as an international and cosmopolitan enterprise. Indeed, the distinguished sociologist Robert Merton has argued that an ethos of universalism is a distinguishing feature of science.[58] More direct was Ernest Rutherford's dictum: 'science is

[53] Elman, *A cultural history*, pp. 200, 198   [54] Elman, *A cultural history*, p. 224
[55] Porter, *Trust in numbers*, p. 25   [56] Kirby, 'Technocratic organisation', pp. 25, 29
[57] Miller, *Science and dissent*, p. 251   [58] Moore, *Disrupting science*, p. 70

international, and long may it remain so'.[59] Early modern Europe had
valued the concept of a Republic of Letters that transcended the interests
of particular states. Yet science, as we have seen, largely emerged in
partnership with the state, and when transported in new realms it was
linked to the emerging state. Indeed, scientific modes of organisation and
application to economic growth promoted the development of the state
where science took root. Balancing the national and the international
elements of science has been a part of its history (as we saw in relation
to the Napoleonic Wars in Chapter 5) and will remain a source of tension.

Before the two world wars, the international dimension of science
was acknowledged through a host of agreements and organisations
that provided a link between nations on the basis of shared scientific
views about such basics as appropriate standards and measurements. The first International Electrical Congress held in 1881 in
Paris, for example, attracted 250 delegates and agreed on using
measurements in terms of ohm, volt and ampere. Further such
congresses were held in Chicago (1893), St Louis (1904) and
London (1908).[60] Since the time of the observation of the transits
of Venus in 1761 and 1769, nations had also worked together on
projects that transcended the reach and resources of a single nation.
During the International Polar Years of 1882–3, for example, eleven
nations sent expeditions supported by the observatories of thirty-five
countries.[61]

Such international cooperation provided the impetus for larger, more
inclusive meetings of scientists. The largest of these was the International
Association of Academies, which first met in Paris in 1900. International
disciplinary meetings also gained ground with, for example, the
International Solvay Conference in Physics in Brussels in 1911 (the
name deriving from the Belgian businessman, Ernest Solvay, who
initiated the conference). Increasing numbers of disciplines were to be
represented by international umbrella bodies, following such examples as
the International Association of Chemical Societies founded in 1911.
The readiness with which these associations were formed underlined
the extent to which it was assumed that all participants from whatever
part of the world would share the same general assumptions about the
fundamentals of their discipline and that there was enough common
ground to ensure fruitful discussion. Internationalism, indeed, was, as
Elisabeth Crawford argues, part of the ethos of science in the period
preceding World War I.[62]

[59] Clark, *The rise of the boffins*, p. 253   [60] Cahan, *Institute*, p. 13
[61] Rose, *Science and society*, pp. 182–3   [62] Crawford, 'The universe', p. 254

World War I shattered this cosmopolitan scientific spirit, leaving international associations in hibernation and necessitating the cancellation of planned conferences. Even in distant Shanghai, French and German scientists no longer would work with each other.[63] Early in the war, in October 1914, anger at Germany, and hence the impulse to exclude it from international discussions, was heightened by an ill-judged open letter with ninety-three signatories including some prominent scientists. It was addressed to 'the civilised world' and amounted to a defence of German militarism, denying reports, for example, of atrocities against civilians in Belgium. Such a manifesto seemed to confirm that German science was no longer truly international and had come under the sway of the state. In short, German science had become 'national science'.[64] Scientific internationalism, then, remained largely in abeyance during the war with the partial exception of journals from both sides being distributed through neutral countries.[65] As war's end approached, there were moves among the Allies to revive the international contact that was considered intrinsic to science. The proposed form combined both national and international dimensions with an International Research Council that was made up of the various National Research Councils.

Membership of the International Research Council, which came into being in 1919, was restricted, however, to the wartime Allies and neutral nations, thus excluding Germany and Austria; Russia, which had abandoned the Allied cause, was also excluded. Germany, it was claimed, was seeking dominance in the sciences just as it had aimed at hegemony through warfare.[66] As the postwar international climate cooled, there were moves to repeal the provision that Germany could not join, and this was finally enacted in 1926. There were subsequent attempts to persuade Germany to join, but these were repulsed by an embittered Germany.[67] Once Hitler came to power in 1933, any prospect of international collaboration with Germany collapsed. Similarly, Stalin's dominance within the Soviet Union also undermined scientific contact outside the country in the 1930s. Tellingly, the International Research Council came to an end in 1931, though another body, the International Council for Scientific Unions, took its place. This viewed itself as the successor body both to the International Research Council (1919–31) and the earlier International Association of Academies (1899–1914). Despite the unpromising conditions of the 1930s and 1940s, it has

---

[63] Osborne, *Nature, exotic*, p. 27   [64] Rasmussen, 'Science and technology', pp. 316–17
[65] Richards, 'Great Britain', p. 184
[66] Schroeder-Gudehaus, 'Nationalism and internationalism', p. 914
[67] Cock, 'Chauvinism and internationalism', pp. 249–50

survived until today, being renamed the International Council for Science in 1998.

Some measure of international collaboration between the wars was maintained by the Rockefeller Foundation, the grants of which encompassed much of the globe and a wide range of European countries (including Germany). Behind the award of such grants was the belief that science was a stimulus to progress that would benefit all nations.[68] Within the Foundation there were different orientations, with the Rockefeller General Education Board being more concerned with institution building in the USA than the International Education Board, which was more open to supporting international research projects.[69] The continuing award of Nobel Prizes also gave science a reward system that transcended particular nations. The award of the Nobel Prize for Chemistry in 1918 to Fritz Haber, the architect of German gas warfare, underlined forcibly the gulf between political and scientific standing. Haber himself saw a clear demarcation between the national and international loyalties of a scientist: 'In wartime, the scholar belongs to his nation, in peacetime to mankind.'[70] The nature of science in the interwar period leant itself to international connections with rapidly changing fields that were being opened up by scientists working in different countries – this was true, for example, of nuclear physics.[71] Even where formal international contact of German scientists was blocked, since they did not belong to the International Research Council, informal ties maintained some semblance of the international spirit in science.[72]

The period between the wars was a test case, then, of the opposing forces of nationalism and cosmopolitanism in science. Nationalism in some cases, such as in the totalitarian regimes of Germany and the Soviet Union, overrode all other loyalties, including that to the international scientific world. What harm this did to science is difficult to measure, but it may account for some of the difficulties with postwar science in Germany and the Soviet Union – nations which were reliant on a generation trained without the benefit of international connections. Internationalism is intrinsic to the way that science works. It made possible scientific competition between nations as they vied for scientific eminence and due acknowledgement from other nations. Without international dissemination of results, too, it is difficult to authenticate results which should be capable of being reproduced in foreign laboratories with quite different political and social systems.[73] But, as this book has shown,

[68] Hamblin, 'Visions', p. 398   [69] Kohler, *Partners in science*, pp. 216, 231
[70] Crawford, 'The universe of international science', pp. 252, 264
[71] Crawford, 'The universe of international science', p. 263
[72] Walker, *German national socialism*, p. 7   [73] Crosland, *Science under control*, p. 434

science has been increasingly a product of the nation state using its resources for the training and support of its scientists. States, then, are reluctant to disseminate results which might be of economic or military advantage to others. Often the problem is solved by the fact that the research has taken place in an industrial or military setting where results are not published. Ultimately, however, all of science is interconnected and excessive secrecy undermines the possibility of scientific stimulation and advance. The conflicting demands of nationalism and internationalism in science admit of no easy balance.

Science, then, needs to operate in both a national and international arena. Historically, the conflict of loyalties thus induced has waxed and waned according to the condition of the larger international order. When governments have been fearful that the findings of the scientific community of their state might give other nations too much advantage, access has been limited. Some governments have feared, too, the very cosmopolitanism of science as a possible source of dissent. Generally, it takes conditions as extreme as war to institute a complete ban on access, since scientists are quick to remind governments that international exchange is in the interests of all. Science is so much a part of the international forum that participation in it forms part of the state building of new nations. Creating a new state by breaking off from an empire or by reshaping an existing polity involves acceptance by the international community. One form of acceptance is involvement in international scientific forums. Such participation also helps consolidate the role of science within the state, thus strengthening the state's control over its population and resources. State-directed scientific improvement in medicine, industry or elsewhere provides an important way in which a state can claim legitimacy and enlist the loyalty of its citizens. Science's international and national dimensions will continue to interact. The sovereign state, however, retains the ultimate power to determine the balance between them.

Epilogue

In the continuing history of the relations between science and the state, 1945 serves as a major dividing line. Before then, science had been important to the state and the state had been important to science, but the partnership between them had been episodic, sometimes tentative and subject to fluctuation. After 1945, the experience of the war and the devastating power of the atomic bomb ensured that science could no longer be consigned to a subordinate role by the state, and that the concerns of scientists would play an integral part in the conduct of government. This was particularly true in the post-World War II super-power, the United States, the attention of which to science had a global influence. The pivotal year 1945 also saw the publication of Vannevar Bush's *Science – the endless frontier*. Its recommendations were only very partially achieved, but it set the terms for discussion about the place of science that shaped debate in the United States and beyond. Above all, it made the case for the importance of research as a worthy recipient of government support. To grasp fully the significance of the end of World War II in the overall dynamic between science and the state we need to look forward as well as backwards. Hence this overview of some of the more significant developments after 1945, to provide further perspective on the extent of change ushered in by the end of World War II.

United States

Since it was the United States that set the pace for the linking of science and the state, it is appropriate to begin by reviewing some of the developments that helped to consolidate this linkage. The wartime scientific organisational structures such as the OSRD were regarded as temporary expedients to be dismantled when war ended. The fear of Vannevar Bush and others was that, as happened after World War I, American society would revert to the way things were before the war without a definite and ongoing commitment to the support of scientific research. It was such a concern that helped prompt *Science – the endless frontier*. What Bush had hoped for was a single agency, the National Science Foundation (NSF), which would coordinate work in all fields of scientific research, including those which he thought would have greatest political appeal: defence and medicine.

Yet, as we saw in Chapter 7, the NSF was not approved by Congress until 1950. The period 1945–50 was not one where the United States' science policy could be left in abeyance. The beginnings of the Cold War, which can perhaps be dated from Winston Churchill's 'Iron Curtain' speech in the United States in May 1946, reaffirmed US determination to maintain its scientific advantage over the USSR. Rather than have to sustain a conventional army the size of that of the USSR, the United States sought to rely on the atomic bomb together with whatever other weapons the partnership of science and the state could produce. Wars of the future, it was clear, would have to be fought with weapons available at the outbreak of hostilities, since there would not be time to devise new weapons in the course of the war, as happened in World War II.[1] Defence research, then, could not wait for the foundation of the NSF, even though it had been Bush's hope that an overarching NSF could give scientists some degree of independence from the military. After various bureaucratic metamorphoses, in 1949 the vast Department of Defense emerged, controlling all aspects of defence. It was to absorb some 80 per cent of government research funds.[2]

Science – the endless frontier's other great drawcard for government support for science had been its role in advancing medical research and treatment. This, however, was to be split off from the proposed NSF. Wartime experience through the Committee on Medical Research of the OSRD helped pave the way for an increase in the size of the National Institute of Health (NIH). Hence the passage of the Public Health Service Act in 1944, with the National Cancer Institute (founded 1937) coming under the National Institute of Health. By 1950, there was a change from Institute to Institutes of Health, a sign of growth and plans for future development.[3]

Wartime experience and the increasing use of scientific expertise, both in military and civilian life, did, however, leave room for some form of agency to support basic research. Before the establishment of the NSF this role was performed by an agency of the navy, traditionally an institution open to scientific innovation. The navy, still smarting from having been excluded from the Manhattan Project and hopeful of developing

---

[1] Roland, 'Science, technology and war', p. 565
[2] Agar, *Science in the twentieth century*, p. 306   [3] Appel, *Shaping biology*, p. 332

nuclear propulsion of ships, established in 1946 the Office of Naval Research (ONR).[4] In the vacuum created by the ongoing debates about the form of the NSF, the Office of Naval Research expanded its role to act as a sponsor of basic research while distinguishing its work from that of the uniformed navy. In the period between 1946 and 1950, of the $86 million allocated for the ONR's contract research only about 10 per cent went to directly naval projects.[5]

For a time, the ONR's responsibilities extended to oversight of the most central scientific project of all, the development of nuclear power and military technology. This, however, became the province of another agency formed by the dissolution of the OSRD's former wide domain: the Atomic Energy Commission,[6] founded in 1947, with the transfer of control of nuclear research from the military to civilian hands. This was an indication of the hopes that nuclear research would yield peaceful benefits. As Senator Brien McMahon argued in proposing the necessary bill, nuclear energy 'is too big a matter to leave in the hands of the generals'.[7] Yet a great deal of the Atomic Energy Commission's research funds went to military ends, and it was under its auspices that the hydrogen bomb was produced in 1952. In the 1950s, 90 per cent of physics research funding at US universities was dispensed by the Atomic Energy Commission, a great deal of it for military projects.[8]

Even after the NSF was founded in 1950, the ONR continued to be a major sponsor of research until the launch of the Russian Sputnik in 1957 galvanised US attention to science – including more of a budget for the NSF.[9] It also prompted the formation of the National Aeronautics and Space Administration in 1958. Come the 1960s there was increasing opposition to navy-sponsored research being conducted on university campuses, and this, together with growing demands within the navy for the ONR research to be orientated to naval needs, led to the scaling down, though not elimination, of the ONR.[10] Its work as a sponsor of scientific research had, however, gone some way towards implementing Bush's goals for the state support of science. It had also highlighted the possibilities of cooperation between an agency of government and the scientific community.[11] The ONR had also drawn on the lessons learnt by the Rockefeller Foundation and passed some of these on to the NSF.[12]

As the cases of the Department of Defense, the Atomic Energy Commission and the ONR suggest, in the aftermath of war the path

---

[4] Sapolsky, *Science and the navy*, p. 10   [5] Wolfe, *Competing with the Soviets*, p. 26
[6] Sapolsky, *Science and the navy*, p. 53   [7] Lapp, *The new priesthood*, p. 89
[8] Bowler and Morus, *Making modern science*, p. 483
[9] Sapolsky, *Science and the navy*, p. 56   [10] Sapolsky, *Science and the navy*, p. 125
[11] Mukerji, *A fragile power*, p. 56   [12] Appel, *Shaping biology*, p. 2

that the partnership between US science and the state took was strongly military in character. War had led to linkages being formed between government, the military, big business and scientists and their universities. Such linkages were strengthened by the Cold War and the Korean War (1950–3). Patrick McGrath writes of the way that the 1940s facilitated the emergence of what he calls 'scientific militarism'.[13] Such linkages with the military were strengthened by the extent to which scientists followed wartime precedent in being called upon to give expert advice. This made them insiders and participants in the growing linkages between the state and what President Eisenhower, in his farewell address of 1961, was to call 'the military industrial complex'.[14] On the other hand, scientists, however eminent, were not immune from the impact of Cold War rhetoric, as Robert Oppenheimer found when he was stripped of his security pass in 1954. Of those investigated by the federal government on security grounds between 1947 and 1954, over half were scientists.[15] The state had, in many ways, become more dependent on scientists but, ultimately, it was the state which set the terms of the partnership.[16]

The predominance of the military was what Vannevar Bush had hoped to avoid: hence his proposal for an NSF that would have overall superintendence of all forms of scientific research. In the event, however, when the relatively minor role which the NSF came to perform was compared with the vast extent of military-directed scientific research, American postwar scientific research had a strongly military character. There was, of course, nothing new about the state forming a partnership with science to advance its military interests – this was an alliance with a long history. What was new was the extent of scientific dependence on the state, with the nature and complexity of science requiring funding on a scale that could only come from the state. The contribution of the philanthropic foundations had been important but, as state support for science increased, they gradually faded out of the role of patrons of science, returning to the fields of health, education and welfare that they had addressed in their early years.[17] Though the NSF was a small agency when compared with the research undertaken by the Department of Defense and the Atomic Energy Commission, it did retain a degree of independence and willingness to support scientific research in areas with no clear military outcome. The NSF saw it as part of its remit to support basic science, which it defined as 'that type of research which is directed toward the increase of knowledge in science'.[18] This underlines the extent

[13] McGrath, *Scientists, business, and the state*, p. 69
[14] Wang, *American science*, pp. 6–8; Wolfe, *Competing with the Soviets*, p. 23
[15] Moore, *Disrupting science*, p. 5   [16] Schweber, 'The mutual embrace of science', p. 6
[17] Kohler, *Partners in science*, p. 406   [18] Appel, *Shaping biology*, p. 142

to which US government support for the sciences was plural, with different agencies pursuing different ends. The biological sciences, which had been relatively neglected during the war, received support from the NSF (though with medical research largely being conducted through the National Institutes of Health).[19] In dealing with the tension between cosmopolitanism and nationalism in science, the NSF largely favoured cosmopolitanism, allowing grants to non-US citizens if the results contributed to US research. During the McCarthyite scares the NIH refused to take account of unproven communist allegations in distributing its grants[20] – being the only major scientific agency to do so.[21]

What did the state gain from support for the NSF? The returns on military and medical research were evident in terms of the nation's image of itself as the premier superpower and of the practical benefits which medicine brought with it. In the postwar world, however, the prestige of science was such that the knowledge that the US led the world in scientific research was a source of prestige for the nation as well as the scientific community. There was, of course, the view, which Vannevar Bush had done much to cultivate, that research was a good in itself and might lead to unexpected practical outcomes. Something of the same rationale prompted the expansion of the National Academy of Science's activities with greater state support for the activities of its National Research Council.[22]

A common thread between the different scientific agencies erected after World War II was the frequent use of the research contract. One lesson that had been learnt from World War II was that scientists worked best when granted some measure of autonomy, which the system of research contracts developed during the war allowed.[23] Putting scientists into uniform had been shown not to work in World War I. The result was a large body of decentralised research carried on in corporations and universities – a distinctive feature of US scientific research.[24] Universities loomed particularly large in the US system, since they did not have to compete with large research institutes such as the German Max Planck Institutes (as the Kaiser Wilhelm Society became after the war) or the French CNRS. In 1953, the proportion of research funding for universities bestowed by the federal government was 54 per cent; by 1965 this had risen to 73 per cent.[25] The American system of widely dispersed universities going back to the Land Grant colleges set up by the Morrill Act of 1862 ensured ample competition between institutions and incentives for applying for

[19] Appel, *Shaping biology*, p. 18   [20] Appel, *Shaping biology*, pp. 81, 117–20
[21] Wang, *American science*, p. 281
[22] Cochrane, *The National Academy of Sciences*, p. 563
[23] Mukerji, *A fragile power*, pp. 8, 41   [24] Dupré and Lakoff, *Science and the nation*, p. 18
[25] Moore, *Disrupting science*, p. 27

grants. Universities, then, became central both to wartime and postwar military research – a source of protest in the 1960s.[26] Within the overall research system there were national laboratories that were especially prominent in atomic research, but even these carried out contracted research.[27] There were also attempts to make such laboratories operate on principles similar to universities, with competition between them being encouraged along with a measure of decentralisation. As government institutions, however, they faced in a more acute form the clash between scientific internationalism and nationalism, especially when it came to the publication of their results.[28] Overall, then, the US research system was largely decentralised, which went some way towards counteracting the natural trend for an increasingly large scientific establishment to come more under the control of an increasingly big government.[29]

Soviet Union

A war-ravaged USSR looked to science to rebuild its society and, so, after 1945, science and scientists received extra resources.[30] The year 1945 was also a watershed year, since it marked the beginning of the final and most determined phase of the Soviet campaign to build an atomic bomb, which was duly delivered in 1949. Such had been the colossal scale of Russian involvement in World War II that it remained a society still psychologically at war, and such an attitude was strengthened by the Cold War. The USSR and the United States became mirror images of each other, both constructing an economy and society built for war. And not a war as of old, but rather one that depended on the quality rather than the quantity of weapons – weapons which needed to be ready as soon as hostilities broke out rather than, as previously, having the possibility of being developed during the course of the war.[31] Cold War science and the armaments built with its aid became imbued with ideological meaning as instruments in the struggle between the two superpowers.[32]

The importance of science brought benefits to scientists, but also a degree of attention from the state, which could interfere with the conduct of research. The end of the war was accompanied by a revival of Stalin's determination to claim the right to determine which form of science best conformed with orthodox Marxism. Hence his renewed support for Lysenko's anti-Mendelian views and persecution of Mendelian geneticists.

[26] Smith, 'The United States', p. 36; Moore, *Disrupting science*, p. 50
[27] Penick et al., *The politics of American science*, p. 27
[28] Westsick, *The national labs*, pp. 1, 18, 22   [29] Lapp, *The new priesthood*, p. 181
[30] Josephson, 'The political economy', p. 154
[31] Roland, 'Science, technology and war', p. 565   [32] Krementsov, *Stalinist science*, p. 93

As we have seen in Chapter 7, Lysenko's high point came in 1948, the year of the Berlin blockade, with Stalin's endorsement of his views going so far as to help revise Lysenko's address.[33] For Lysenko's followers and, no doubt, for Stalin himself, what was at stake was the uniqueness of Soviet science as against Western science, a reflection of the Cold War ideological struggle. Hence the remark of one of Lysenko's followers that 'there is and can be nothing in common between our science and so-called "world science"'.[34] Ideology, however, was liable to prove more flexible if it impeded progress in developing weapons to counter the USSR's Cold War enemies. While the end of the war saw something of a revival of the campaign against those aspects of modern physics that did not conform to a materialist model of the world, this abated as it became clear that such forms of physics were necessary to develop new atomic weapons.[35]

Though the scale of Soviet science expanded postwar, the structures which sustained it did not greatly change from the ones used in the war. This perhaps reflects the fact that the central state was the only source of support for science, so the need for a diversity of funding bodies was less of a concern. The Academy got bigger with more institutes, but new forms of state reorganisation of the conduct of science were not established as they were in the United States. There was an awareness that over-centralisation had been a liability during the war, and hence there were moves to provide greater geographical dispersal of the Soviet science establishment. A whole new research city, 'Academy City', for example, was built in Siberia in 1958.[36] There was also an attempt to diversify on gender grounds, with the USSR going much further than other countries in opening up scientific positions to women. Women eventually held about one half of Academy positions thanks to state policy, though they tended to be concentrated in the lower ranks.[37] By the early 1960s the forces of centralisation were again uppermost, and in 1961 a State Committee of the Council of Ministers for the Coordination of Scientific Research (from 1965 the State Committee for Science and Technology) was established and was to last until 1991. This focussed on applied research while the Academy was more concerned with basic research.[38] Part of this demarcation was the restructuring of the Academy so that it lost about half of its institutes, which were then absorbed by industrial ministries.[39]

So deeply entrenched was the Soviet system of state support for science with the Academy at its head that it had difficulty coming to terms with the post-Soviet era following the fall of the Berlin Wall in 1991.

[33] Krementsov, *Stalinist science*, p. 180   [34] Krementsov, *Stalinist science*, p. 179
[35] Krementsov, *Stalinist science*, p. 282   [36] Vucinich, *Empire of knowledge*, p. 285
[37] Vucinich, *Empire of knowledge*, p. 293
[38] Vucinich, *Empire of knowledge*, pp. 302, 308   [39] Graham, *Science in Russia*, p. 184

The Soviet Academy became the Russian Academy, with the former republics now with their separate academies. State control over the Academy was still in evidence with the president of the Academy having to be ratified by the president of the Russian federation. The Academy also had to deal with the science budget being reduced to about 20 per cent of what it had been before 1991. This situation was slightly alleviated by grants from foreign foundations, but they brought with them ways of selecting fundable projects such as open competition that were foreign to Russian practice. Part of the difficulty in post-Soviet Russia is that science no longer has the same ideological force as it did when Marxism was the state creed.[40] Gone are the days from about the 1960s when the Soviet Union spent a greater proportion of its national income on science than any other country, including the United States.[41]

Britain

The year 1945 was a watershed year for British science policy as it was for the United States, the USSR and other nations. The extent of change, however, was not as extensive as in the United States, which built up a whole new scientific infrastructure as a result of its wartime experience that was consolidated by the Cold War. Much of Britain's scientific infrastructure went back to World War I and had not been changed radically during the war as it had been in the United States. Nonetheless, postwar there were some significant changes, with separate Advisory Committees being set up in 1947 to cover both defence and military research – these both being successors of the Science Advisory Committee that had existed during the war. In the preceding year, a Department of Defence was established, thus centralising the work that had previously been conducted by the three services. Military research predominated as it did in the United States, with the largest British postwar research project being the production of the atomic bomb in 1952, followed by the hydrogen bomb in 1957. Both reflected British determination to have an independent deterrent in the face of uncertainty about US willingness to involve itself in conflict in Europe. In contrast to the US policy of contracting out as much research as possible, the British used in-house institutions for defence research (many run by the Department of Supply) together with private armament companies.[42]

But civilian as well as defence-related research was increased. In 1950, the Colonial Research Office was established with an annual turnover of

---

[40] Graham and Dezhina, *Science in the new Russia*, pp. vii, ix, 18
[41] Fortescue, *Science policy*, p. 1   [42] Edgerton, 'Science in the United Kingdom', p. 5

one million pounds a year, making it the second-largest civilian agency for promoting research after the DSIR.[43] The DSIR remained the central institution for civilian research until 1964, when it was abolished. This reflected not a diminution of the British state's interest in supporting science but rather the opposite. It was thought that other structures could better initiate new fields than the DSIR, which was essentially a coordinating body.[44]

Confidence that science could generate wealth remained strong, reflecting the power that science had exercised in the war. The British Labour Prime Minister, Harold Wilson, went so far as to claim that 'if there is one word I would use to identify modern socialism it is 'science'".[45] (Such confidence was, however, to fade from about the 1970s.)[46] To strengthen governmental support for science, the DSIR was replaced by three new research councils: the Science Research Council, the Natural Environmental Research Council and the Social Sciences Research Council, supplementing the long-existing councils for medicine and agriculture. As the medical research council had shown, the council structure provided an extra layer of protection in keeping the scientists from too much control by the state. An increasing amount of research was conducted in universities, the research capabilities of which had expanded. Thus, there was some movement towards the US decentralised model of research, in contrast to the more centralised model embodied in the DSIR and in the short-lived Ministry of Science founded in 1959. Military research was not, however, conducted on any scale in British universities in contrast to the United States.[47]

Germany

Plainly, for Germany, 1945 marked a decisive break, with the fall of the Nazi regime and the occupation of the country by the Allies. Amidst all the confusion there was considerable concern with the rebuilding of German science. This was perhaps because, as was the case after World War I, it was viewed as demonstrating German cultural power at a time when it had been decisively defeated militarily. The adaptable KW Society began planning its transition to the new regime before the war had finished, moving its headquarters from Berlin to Göttingen to be out of range of the advancing Russian armies. Its rationale for continued existence and past compliance with the Nazis was that it was a private,

---

[43] Clarke, 'The chance to send', p. 188   [44] Varcoe, *Organising for science*, p. 78
[45] Rose and Rose, *Science and society*, p. 93   [46] Wilkie, *British science and politics*, p. 19
[47] Edgerton, 'British scientific intellectuals', p. 5

nongovernmental body. Nonetheless, it sought assistance from the postwar state and, once a new German state had been established, received a generous allocation from both the federal and state governments.[48] As part of adjusting to the postwar world it changed its name in 1948, at the insistence of the British, to the Max Planck Institute for the Advancement of Society. Another point of continuity with the imperial past was the continued existence of the Reichsanstalt (Imperial Institute) once it had been shorn of its 'imperial' designation and renamed as the Physikalisch-Technische Bundesanstalt (Federal Physical and Technical Institution).[49]

Despite the travails of postwar Germany, further state funding followed using the model that had been employed after World War I of the Notgemeinshaft (Emergency Fund)[50] – the very name an indication of the renewed fears that German science might lose its eminence. This was a temporary measure, as the Emergency Fund was absorbed into the Deutsche Forschungsgemeinschaft (German Research Council) in 1951. This has provided the structure for a state-supported scientific establishment organised along federal and decentralised lines, in contrast to the increasing centralisation of the Nazis. Government funds are distributed through three main avenues: university institutes, Max Planck institutes and 'Big Science' state institutes. By contrast, East Germany adopted the Soviet centralised academy model of supporting science, until the regime collapsed.[51] For Germany, 1945 was the beginning of a new regime in the state's relations with science with an amalgam of the old and the new. In the rebuilding of Germany support of science was given a high priority, an indication of how strong was and is the linkage between the German state and science.

France

For France, 1945 was less of a watershed than other major Western countries. Early defeat in World War II left intact most of its institutions, including those intended for the support of science. The challenge in the postwar world was to use such institutions once again to project France's eminence in science and a major place in world affairs generally. In this the state played a central role. De Gaulle encapsulated the close nexus between science and the French state with his remark that 'scientific research has become one of the means of state policy'.[52] The major

[48] Schüring, 'Expulsion, compensation, and the legacy', p. 310
[49] Agar, *Science in the twentieth century*, pp. 495–6
[50] Deichmann and Muller-Hill, 'Biological research', p. 182
[51] Walker, 'Twentieth-century German science', pp. 816–17   [52] Gilpin, *France*, p. 441

instrument for keeping centralised control over science was the research funding institution, the CNRS, founded just before World War II but achieving its greatest impact in the postwar world. Control over funding meant that the state could determine scientific priorities, which gave greater weight to applied science and engineering than to basic science.

Centralisation brought the familiar mixture of advantages and disadvantages. French science developed a coherence and overall design that set it apart from other Western countries. Indeed, it had more in common with more recently industrialised countries such as Japan. On the other hand, the rigidities of the system made major change, such as the founding of a new institute, difficult. The centralised model also made it difficult to combine research with teaching at a university level so that in France, to a large extent, teaching and research occur in different institutions. Science, then, has almost been absorbed by the state so that the scientific community has almost become a part of the structures of the state.[53] This has reduced autonomy, but has increased the prestige and resources of science in a nation that has looked on science as part of the same rational frame of mind as led to the creation of the Republic.

Japan

Both France and Japan share a planned scientific estate, but the role of the state varies considerably. In France, the state looms large, to the point where it almost subsumes the scientific community. In Japan, on the other hand, a great deal of scientific activity is carried on in the private sector. Against this, however, the divide between the public and private sector in Japan is not always clear cut, with business and the governmental bureaucracy often in close alliance.[54] This applies particularly to the Ministry for International Trade and Industry (MITI), founded in 1949 to help restore the Japanese economy after wartime devastation. Along with its numerous contacts with business it also has its own in-house laboratories.

The Japanese state does have direct control over some sectors of the scientific community, particularly those in the universities – even though the German-based centralised university system was changed after the war by the Americans to a more decentralised one. The end of the war marked a watershed in Japanese science in other ways.[55] The prewar Imperial Academy and National Research Council, which acted as intermediaries between the state and the scientific community, were abolished

[53] Baumgartner and Wilsford, 'France: science within the state', pp. 63
[54] Low, Nakayama, Yoshioka, Science, technology and society, pp. 118–19
[55] Nakayama, Science, technology and society, pp. 47–8

and replaced in 1949 by the Japan Science Council. This, unusually, is elected by the entire Japanese science community and is intended to represent it. Along with advising government on scientific issues, it distributes the government research budget.[56] Another postwar instrumentality was the Science and Technology Agency, founded in 1954 to facilitate Big Science projects such as nuclear energy.[57] The limited size and influence of such bodies reflects the nature of the Japanese economy, which has one of the lowest government sectors among industrialised nations – thus, much is left to private enterprise. Predictably, private sponsorship of science has resulted in relative neglect of basic science, though there have been attempts to remedy this.[58]

China

To Mao Zedong, victory in 1949 and the establishment of the People's Republic of China meant establishing a new order based on the Marxist-Leninist principles that had been transmitted from the Soviet Union. These included a high regard for science as a confirmation of materialism. It also included the view that the Party had the right to determine what forms of science were ideologically sound. For Mao, the mark of proletarian science was that it contributed to the welfare of the common people. Science that was too theoretical he regarded with suspicion. Hence, like Stalin, he attacked branches of modern physics which appeared to stray from the material models – Einstein's theories were a particular target of criticism during the Cultural Revolution.[59] From the Soviets, too, he derived an adulation for the work of Lysenko with his closeness to the work of peasant farmers. In China, however, though Mendelian genetics was attacked, its exponents were not subject to persecution.[60] Another legacy from Russia was the academy model of scientific research. Under Mao, the Chinese Academy of Sciences oversaw most scientific research, while the universities were largely stripped of their research functions.[61]

As part of the economic modernisation of China, Mao's successor, Deng Xiaoping (paramount leader of China from 1978 to 1989) attempted to develop a less rigidly controlled system that would even leave some room for market forces[62] – in so far as this did not weaken the power of the Party. The Academy's near monopoly was weakened, and the universities began

[56] Long, 'Policy and politics', p. 426
[57] Nakayama, *Science, technology and society*, p. 107
[58] Frieman, 'People's Republic of China', p. 136
[59] Fan, 'Science, state, and citizens', p. 235   [60] Schneider, 'Learning from Russia', p. 46
[61] Simon and Goldman, 'The onset', p. 7   [62] Saich, 'Reform of China's science', p. 70

to regain the research functions which they had enjoyed before World War II. Ideological sanctions against modern physics were eliminated and, conspicuously, there was an ostentatious ceremony to mark the anniversary of Einstein's birth in 1979.[63] The Party, however, continued to keep control of science to ensure that it was primarily orientated to utilitarian ends. In the wake of the Tiananmen Square uprising (1989), the scientific community came under some suspicion for its associations with Western liberalism,[64] but this did not lead to major change. One of the most significant features of Chinese science since Mao has been its openness to contact with the West, with a major programme for allowing foreign scientists into China and, conversely, allowing Chinese scientists abroad. From 1980 to 1989, 100,000 students went abroad, 75 per cent of these to the USA.[65] Post-Mao Chinese science, then, has embraced the international dimension of science while also endeavouring to build up a national scientific establishment.

India

The newly created Indian state, which achieved independence in 1947, was one in which science formed part of the national ideology. As we have seen in Chapter 8, Jawaharlal Nehru, the first prime minister, saw science as a force which would liberate India from the heritage of imperialism as well as the hold of tradition.[66] So important was science to Nehru's conception of nation building that, along with the prime ministership, he held the first position of minister of science and technology. Nehru's aim, as he proclaimed when opening the National Institute of Sciences in 1948, was to create a society where 'actions are governed by a scientific approach'.[67] The state was to be his means of achieving such goals, building on an imperial legacy which had favoured state research institutes rather than universities.[68] Hence, he established a number of state-directed institutions to promote science and technology, such as the Saha Institute of Nuclear Physics (f. 1950) and the Tata Institute of Fundamental Research (1946).[69]

The Postwar Scientific Settlement

For India, as for China, another newly industrialising nation, science was closely linked to the central structures of a state committed to rapid development. While India still encased science within the structures of

---

[63] Agar, *Science in the twentieth century*, p. 427   [64] Miller, *Science and dissent*, p. 236
[65] Frieman, 'People's Republic of China', p. 139   [66] Arnold, 'Nehruvian science', p. 366
[67] Harrison and Johnson, 'Introduction', p. 1   [68] Arnold, 'Nehruvian science', p. 366
[69] Agar, *Science in the twentieth century*, p. 511

a democratic state, in China the Chinese Academy of Sciences acted very largely as a servant of the Party. The forms that the partnership between science and the state might take after World War II were, then, various. The United States built on its wartime legacy to build up a system based around the research grant and research contract, which lessened the direct control of the state by producing a largely decentralised and pluralistic scientific establishment. Nonetheless, the high level of state expenditure tilted the system towards state ends, particularly military ones. The Soviet Union, having been victorious in the titanic struggle of World War II, largely continued with the structures in place then. In particular, the Soviet Academy of Sciences with its many institutes was elevated to a position approximating a department of state. Britain continued with a structure that largely derived from World War I, but which was restructured in gradual stages after World War II to take science out of the direct control of a government department. Germany maintained a surprising degree of continuity with its prewar state scientific apparatus, albeit with such superficial changes as the renaming of the Kaiser Wilhelm Society as the Max Planck Institute. Despite its claims to be a private institution, the Max Planck Institute was largely sustained, at least in the immediate postwar period, by government. The formation of research councils, however, reduced the direct control of the state. True to its traditions of centralisation, the French state incorporated science into its larger structures of government, giving science an assured basis of support and prestige but limiting its autonomy. Like France, Japan has also incorporated science into its larger goals for society, though much of the research has been carried out in a private sector that is closely linked with the centralised bureaucratic structures concerned with planning. At the time of its independence, India drew science into the structures of a socialist state, though this has loosened with time as the founding socialist ideals have faded.

Postwar state partnership with science, then, has taken many forms, but what is common to them all is a recognition of the importance of science as a part of the state structure. After Hiroshima, no one could doubt the awesome power that science could unleash, nor that the state needed to make some rapprochement with science. State legitimacy depended on displaying to its population that the state had taken advantage of the benefits that science could bestow in fields as various as economics, medicine and defence. The post-World War II world, then, made manifest the extent to which science and the state were bound together in a reciprocal relationship. This involved the state providing hitherto unheard-of levels of support in return for the kudos and social benefits science conferred on the state. So uncontested was the role of science that the postwar period has seen a move away from the scientific nationalism that was growing in the interwar period. The international dimension of science is now generally seen as of benefit to all so long as it does not infringe on state secrets. Indeed, influential figures in the United States such as President Eisenhower have linked openness in science to a more general internationalism in US foreign policy.[70]

Conclusion

Prestige had always been an important part of the science-state relationship, with the marginal activity of science enjoying court support in the time of the Scientific Revolution because of the aura of supporting the discovery of new and often secret knowledge. As state structures took root and predictable bureaucratic routines were established, science also served to confirm the virtues of governmental rationality. Science in the form of 'political arithmetic' and early demography provided tools by which populations could be governed and their needs predicted. This was taken still further by the Enlightenment-tinged cameralists of the German lands in the eighteenth century. Science was also, of course, supported by the state for the practical benefits it could confer, though these took time to emerge. Astronomy was one area where science plainly conferred practical benefits through its use in navigation, hence the erection of royal observatories in early modern Europe.

The importance of science in warfare was also slow to emerge, though the French chemists of the Revolutionary period played a part in achieving victory. The dynamics of the science-state relationship changed remarkably with the 'scientisation' of industry, the merging of technology with science in the latter nineteenth century. This was accompanied by the increasing readiness of states to support science for the economic benefits it could bring. The scale of industry also made science an important partner in war, a linkage with the state. This accelerated the involvement of the state over the course of the twentieth century to the point where, since World War II, no state could afford to overlook the importance of science. The modern state and science now form an inseparable bond that is embodied in the structures of government.

What does the changing relationship between science and the state reveal about the nature of science? Science has proved remarkably adaptable. Many different forms of state have been involved in the sponsorship of science and science has operated under them all. Scientists continued with their work under totalitarian governments as well as liberal democracies.

---

[70] Hamblin, 'Visions', p. 394

In some systems scientific activity has been closely controlled from the centre, while in others science has been granted freer rein. Inevitably, government money brings with it greater control, and the increasing importance of science to government has brought with it greater accountability. The forms which this accountability can take vary: in the Soviet Union, science was meant to be planned in the same way that the economy was with predictions of what would be achieved in a prescribed period. In the United States, where research largely takes place in universities rather than in institutes, grants are generally made to an individual researcher or team that faces sanctions both from their university and the grant-giving body for not producing.

Some institutions seem more effective than others at promoting scientific activity, though this can vary over time. The universities were initially relatively impervious to the Scientific Revolution in early modern Europe, but in nineteenth-century Germany they were at the heart of a research culture that linked science with industry. So, too, in twentieth-century America the university has been the main partner of state-sponsored research bodies, with remarkable success in both war and peace. The French Academy of Science in the eighteenth and early nineteenth centuries led Europe in the promotion of new branches of science. State-sponsored scientific institutes, however, have fared less well in the twentieth century, since they have often proved much less nimble than a system of grants to individuals in embracing change. Block grants to institutes often solidify the status quo, which is further strengthened by the shared assumptions of others in the institute; a university department, by contrast, will include others from other fields and lines of approach.[71] The state's choice of scientific infrastructure can, then, prompt different forms of scientific development. Where states bestow their money is, of course, also a major factor in building up some fields as against others. The state, then, can help shape the nature of science, but so too can science help shape the nature of the state. The modern state embodies the truth of Bacon's adage about the relationship between knowledge and power. The forms of knowledge which science has made possible have helped to make possible the workings of the modern state. The state, too, derives part of its prestige, both domestically and internationally, from its level of scientific achievement. Bacon's intimation that both science and the state could form a fruitful partnership has come to fruition with the passage of the centuries.

---

[71] Gustafson, 'Why doesn't Soviet science', p. 35

Bibliography

Abelson, Philip H., 'Knowledge and power: science in the service of society' in Lakoff (ed.), *Knowledge and Power*, pp. 469–82.

Agar, Jon, *Science in the twentieth century and beyond*, Cambridge: Polity Press, 2012.

Albrecht, Ulrich, 'Military technology and National Socialist ideology' in Renneberg and Walker (eds.), *Science, technology and National Socialism*, pp. 88–125.

Alder, Ken, 'A revolution to measure: the political economy of the metric system in France' in M. N. Wise (ed.), *Values of precision*, Princeton: Princeton University Press, 1995, pp. 39–71.

  Engineering the Revolution: Arms and Enlightenment in France, 1763–1815, Chicago: Chicago University Press, 1997.

Allen, Douglas W., *The institutional revolution: Measurement and the economic emergence of the modern world*, Chicago: Chicago University Press, 2011.

Alter, Peter, *The reluctant patron: science and the state in Britain, 1850–1920*, Oxford: Berg, 1987.

Anderson, Benedict, *Imagined communities: reflections on the origins and spread of nationalism*, London: Verso, 1991.

Andrews, James T., *Science for the masses: The Bolshevik state, public science, and the popular imagination in Soviet Russia, 1917–1934*, College Station: Texas A&M University, 2003.

Anon, *Baconiana Bibliographica: or certain remains of the Lord Bacon concerning his writings*, London, 1679.

Appel, Toby, *Shaping Biology: The National Science Foundation and American Biological Research, 1945–1975*, Baltimore: Johns Hopkins University Press, 2000.

Arnold, David, *The new Cambridge history of India*, vol. III (5):*Science, technology and medicine in colonial India*, Cambridge: Cambridge University Press, 2000.

  'Nehruvian science and postcolonial India', *Isis*, 104, 2013, 360–70.

Auerbach, Lewis E., 'Scientists in the new deal: a pre-war episode in the relations between science and government in the United States', *Minerva*, 3 (4), 1965, 457–82.

Azzolini, Monica, *The duke and the stars: astrology and politics in Renaissance Milan*, Boston: Harvard University Press, 2012.

Baber, Zaheer, *The science of empire: Scientific knowledge, civilization, and colonial rule in India*, Albany: Albany State University of New York Press, 1996.

Bacon, Francis, *The works of Francis Bacon*, (eds.) James Spedding, Robert Leslie Ellis and Douglas Denon Heath, Boston: Brown and Taggard, 1861–4, 1900.
   The advancement of learning and the new Atlantis, London: Oxford University Press, 1951.
Baker, Keith Michael (ed.), *Condorcet: Selected writings*, Indianapolis: Bobbs-Merrill, 1976.
Barrera-Osorio, Antonio, *Experiencing nature: the Spanish American empire and the early scientific revolution*, Austin: University of Texas Press, 2006.
   'Knowledge and empiricism in the sixteenth-century Spanish Atlantic world' in Bleichmar, De Vos, Huffine and Sheehan (eds.), *Science*, pp. 219–32.
Bartholomew, James R., *The formation of science in Japan: Building a research tradition*, New Haven: Yale University Press, 1989.
   'Science in twentieth-century Japan' in John Krige and Dominique Pestre (eds.), *Science in the twentieth century*, pp. 839–96.
Basu, Aparna, 'The Indian response to scientific and technical education in the colonial era, 1820–1920' in Kumar (ed.), *Science and Empire*, pp. 126–38.
Baumgartner, Frank and Wilsford, David, 'France: science within the state' in Solingen (ed.), *Scientists and the state*, pp. 63–91.
Baxter, James Phinney, *Scientists against time*, Cambridge, MA: MIT Press, 1952.
Bayly, C. A., *Imperial meridian: the British empire and the world, 1780–1830*, London: Longman, 1989.
Bedini, Silvio A., *Thomas Jefferson: Statesman of science*, New York: Macmillan, 1990.
Bennett, Brett M. and Hodge, Joseph M. (eds.), *Science and empire: Knowledge and networks of science across the British Empire, 1800–1970*, Houndmills: Palgrave, 2011.
Ben-David, Joseph, *Centers of Learning: Britain, France, Germany, United States*, New York: McGraw-Hill, 1977.
Benson, Keith R., 'Field stations and surveys' in Bowler and Pickstone (eds.), *The Cambridge history of science*, vol. 6, pp. 76–89.
Berman, Morris, *Social change and scientific organization: The Royal Institution, 1799–1844*, London: Heinemann, 1978.
Beyerchen, Alan, *Science under Hitler*, New Haven: Yale University Press, 1977.
   'On the stimulation of excellence in Wilhelmian science' in Jack R. Dukes and Joachim Remak (eds.), *Another Germany: A Reconsideration of the Imperial Era*, London: Westview Press, 1988, pp. 139–68.
Beyler, Richard H., 'Maintaining discipline in the Kaiser Wilhelm Society during the National Socialist regime', *Minerva*, 44, 2006, 25–66.
Beyler, Richard, Kojevnikov, Alexei and Wang, Jessica, 'Purges in comparative perspective: rules for exclusion and inclusion in the scientific community under political pressure', *Osiris*, 20, 2005, 23–48.
Biagioli, Mario, 'Scientific Revolution, social bricolage and etiquette' in Roy Porter and Mikuláš Teich (eds.), *The Scientific Revolution*, pp. 11–54.
   Galileo, courtier the practice of science in the culture of absolutism, Chicago: University of Chicago Press, 1993, pp. 11–54.

Bleichmar, Daniela, 'A visible and useful empire. Visual culture and colonial natural history in the eighteenth-century Spanish world' in Daniela Bleichmar, Paul De Vos, Kristin Huffine, Kevin Sheehan (eds.), *Science*, pp. 290–310.

Bleichmar, Daniela, De Vos, Huffine, Paula, Kristin and Sheehan, Kevin (eds.), *Science in the Spanish and Portuguese Empires, 1500–1800*, Stanford: Stanford University Press, 2008.

Bonneuil, Christophe, 'Development as experiment: science and state building in late colonial and postcolonial Africa, 1930–1970', *Osiris*, 15, 2000, 258–81.

Boschiero, Luciano, 'Natural philosophizing inside the late seventeenth-century Tuscan court', *British Journal for the History of Science*, 35 (4), 2002, 383–410.
    Experiment and natural philosophy in seventeenth-century Tuscany. *The history of the Accademia del Cimento*, Dordrecht: Springer, 2007.

Bowler, Peter J. and Morus, Iwan Rhys, *Making modern science: a historical survey*, Chicago: University of Chicago Press, 2005.

Bowler, Peter J. and Pickstone, John V. (eds.), *The Cambridge history of science*, vol. 6. *Modern life and earth sciences*, Cambridge: Cambridge University Press, 2009.

Brewer, John, *The sinews of power: War, money and the English state, 1688-1783*, Cambridge, MA: Harvard University Press, 1990.
    'The eighteenth-century British state. Contexts and issues' in Lawrence Stone (ed.), *An imperial state at war: Britain from 1689 to 1815*, London: Routledge, 1994, pp. 52–71.

Brockway, Lucile H., *Science and colonial expansion: the role of the British Royal Botanic Gardens*, New York: Academic Press, 1979.

Brooks, Nathan M., 'Munitions, the military and chemistry in Russia' in R. MacLeod and J. A. Johnson (eds.), *Frontline and factory*, pp. 75–101.

Bruce, Robert V., *The launching of modern American science 1846–1876*, New York: Alfred A. Knopf, 1987.

Bulmer, Martin, 'Knowledge institutionalized: higher education and philanthropic foundations', *Minerva*, 40, 2002, 189–201.

Burns, William E., *The scientific revolution in global perspective*, Oxford: New York, 2016.

Bush, Vannevar (ed.), *Science – the endless frontier*, Washington, DC: US Office of Scientific Research and Development, 1960.

Cahan, David, 'Institutional revolution in German physics', *Historical Studies in the Physical Sciences*, 15 (2), 1985, 1–65.
    An institute for an empire: the Physikalisch-Technische Reichsanstalt, 1871–1918, Cambridge: Cambridge University Press, 1989.

Cañizares-Esguerra, Jorge, *Nature, empire, and nation: explorations of the history of science in the Iberian world*, Stanford: Stanford University Press, 2006.

Cardwell, D. S. L., *The organisation of science in England*, London: Heinemann, 1972.

Carroll, Patrick, *Science, culture and modern state formation*, Berkeley: University of California Press, 2006,

Chambers, David Wade, 'Period and process in colonial and national science' in Nathan Reingold and Marc Rothenberg (eds.), *Scientific colonialism: a cross-*

cultural comparison, Washington, DC: Smithsonian Institute Press, pp. 297–321.

Chauvreau, Sophie, 'Mobilization and industrial policy: chemicals and pharmaceuticals in the French war effort' in R. MacLeod and J. A. Johnson (eds.), *Frontline and factory*, pp. 21–30.

Clark, Roland W., *The rise of the boffins*, London: Phoenix House, 1962.

Clarke, Sabine, 'Pure science with a practical aim: the meanings of fundamental research in Britain, circa 1916–1950, *Isis*, 101:2, 2010, 285–311.

'"The chance to send their first class men out to the colonies": the making of the colonial research service' in Bennett and Hodge (eds.), *Science and empire*, pp. 187–208.

Close, Charles, *The early years of the Ordnance Survey*, Newton Abbot: David & Charles, 1969.

Coben, Stanley, 'American foundations as patrons of science: the commitment to individual research' in Reingold (ed.), *The sciences in the American context*, pp. 229–48.

Cochrane, Rexmond, *The National Academy of Sciences: The first hundred years 1863–1963*, Washington, DC: National Academy of Sciences, 1978.

Cock, A. G., 'Chauvinism and internationalism in science: The International Research Council, 1919–1926', *Notes and Records of the Royal Society of London*, 37 (2), 1983, 249–88.

Cohn, Bernard, *Colonialism and its forms of knowledge. The British in India*, Princeton: Princeton University Press, 1996.

Coopersmith, Jonathan, 'The role of the military in the electrification of Russia, 1870–1890' in Mendelsohn, Smith and Weingart (eds.), *Science, technology and the military*, II, pp. 291–305.

Cormack, Lesley, 'Twisting the lion's tail: practice and theory at the court of Henry Prince of Wales' in Moran (ed.), *Patronage and institutions*, pp. 67–83.

Crawford, Elisabeth, 'The universe of international science, 1880–1939' in Tore Frängsmyr (ed.), *Solomon's House revisited*, pp. 251–69.

Crosland, Maurice, 'The congress on definitive metric standards, 1798–1799: The first international scientific conference?', *Isis*, 60 (2), 1969, 226–31.

'The French Academy of Sciences in the nineteenth century', *Minerva*, 16 (1), 1978, 73–102.

Science under control the French Academy of Sciences, 1795–1914, Cambridge: Cambridge University Press, 1992.

Cudworth, Erika, Hall, Timothy and Mc Govern, John (eds.), *The modern state. Theories and ideologies*, Edinburgh University Press, 2007.

Daniels, George (ed.), *Nineteenth-century American science. A reappraisal*, Evanston: Northwestern University Press, 1972.

Daston, Lorraine, 'Nationalism and scientific neutrality under Napoleon' in Frängsmyr (ed.), Solomon's house, pp. 95–119.

Day, C. R., 'Education for the industrial world: technical and modern instruction in France under the Third Republic, 1870–1914' in Fox and Weisz (eds.), *The organization of science*, pp. 127–53.

Dean, Warren, 'The green wave of coffee: beginnings of tropical agricultural research in Brazil (1885–1900), *The Hispanic American Historical Review*, 69 (1) 91–115.

Dear, Peter, *Revolutionizing the science: European knowledge and its ambitions, 1500–1700*, Basingstoke: Palgrave, 2001.

De Sauvigny, Guillaume De Bertier, 'Science et politique sous la Restauration', *History and technology*, 5 (2–4), 1988, 273–99.

Deichmann, Ute, *Biologists under Hitler*, Cambridge: Harvard University Press, 1996.

Deichmann, Ute and Muller-Hill, Benno, 'Biological research at universities and Kaiser Wilhelm Institutes in Nazi Germany' in Renneberg and Walker (eds.), *Science, technology and national socialism*, pp. 160–83.

Devles, Daniel, 'On the flaws of American physics: a social and institutional analysis', in George Daniels (ed.), *Nineteenth-century American science*, pp. 133–51.

Dmytryshyn, Basil, Crownhart-Vaughan, E. and Vaughan, T. (eds.), *Russian penetration of the North Pacific Ocean, 1700–1790: a documentary record*, Eugene: Oregon Historical Society Press, 1988.

Dorn, Harold, *The geography of science*, Baltimore: Johns Hopkins, 1991.

Drayton, Richard, *Nature's government: science, imperial Britain, and the 'improvement' of the world*, New Haven: Yale University Press, 2000.

Dunmore, John (ed.), *The journal of Jean-François de Galaup de la Perouse, 1785–1788*, London: Hakluyt Society, 2002.

Dupré, J. Stefan and Lakoff, Sanford A., *Science and the nation. Policy and politics*, Eaglewood Cliffs: Prentice-Hall, 1962.

Dupree, A. Hunter, *Science and the emergence of modern America 1865–1916* Chicago: Rand Mc Nally, 1963.

Science in the federal government: a history of policies and activities to 1940, New York: Harper & Row, 1964.

'The national pattern of American learned societies, 1769–1863' in Alexandra Oleson and Sanborn C. Brown (eds.), *The pursuit of knowledge in the early American Republic: American scientific and learned societies from colonial times to the Civil War*, Baltimore: John Hopkins University Press, 1976, pp. 21–32.

Dyson, Kenneth H. F., *The state tradition in Western Europe: A study of an idea and institution*, Oxford: Martin Robertson, 1980.

Eamon, William, 'Court, academy and printing house: patronage and scientific careers in late Renaissance Italy' in Moran (ed.), *Patronage and institutions*, pp. 23–50.

Eastwood, '"Amplifying the province of the legislature": the flow of information and the English state in the nineteenth century', *Historical research*, 62, 1989, 276–94.

Edgerton, David, 'Science and war' in Olby, Cantor, Christie and Hodge (eds.), *Companion to the history of modern science*, pp. 920–33.

'Science in the United Kingdom. A study of the nationalization of science' in John Krige and Dominique Pestre (eds.), *Science in the twentieth century*, pp. 759–76.

'British scientific intellectuals and the relations of science, technology and war' in Forman and Sánchez-Ron, *National military establishments*, pp. 1–35.

Britain's war machine: weapons, resources, and experts in the Second World War, Oxford: Oxford University Press, 2011.

Elliott, J. H., *Imperial Spain, 1469–1716*, London: Edward Arnold, 1963.

Elman, Benjamin A., *A cultural history of modern science in China*, Cambridge, MA: Harvard University Press, 2006.

Epstein, Joel J., *Francis Bacon: a political biography*, Athens, Ohio: Ohio University Press, 1977.

Eriksson, Gunnar, 'Commentary' in Frängsmyr (ed.), *Solomon's House*, pp. 31–5.

Fan, Fa-ti, *British naturalists in Qing China. Science, empire and cultural encounter*, Cambridge, MA: Harvard University Press, 2004.

'Science, state, and citizens: notes from another shore', *Osiris*, 27 (1), 2012, 227–49.

Findlen, Paula, 'Economy of scientific exchange in early modern Italy' in Moran (ed.), *Patronage and institutions*, pp. 5–24.

Possessing nature: Museums, collecting, and scientific culture in early modern Italy, University of California Press, 1994.

'Cabinets, collections and natural philosophy' in Eliška Fučíková et al. (eds.), *Rudolf II and Prague: the court and the city*, London: Thames and Hudson, 1997, pp. 201–19.

'Anatomy theatres, botanical gardens, and natural history collections' in Park and Daston (eds.), *The Cambridge history of science*, vol. 3, pp. 272–89.

Finn, Stephen J., *Thomas Hobbes and the politics of natural philosophy*, London: Continuum, 2006.

Fischer, Wolfram and Lundgreen, Peter, 'Recruitment and training of administrative and technical personnel' in Charles Tilly (ed.), *The formation of national states*, pp. 456–651.

Foote, George A., 'The place of science in the British reform movement 1830–50', *Isis*, 42 (3), 1951, 192–208.

Forman, Paul, 'Scientific internationalism and the Weimar physicists: the ideology and its manipulation in Germany after World War I', *Isis*, 64 (2), 1973, 150–80.

'The financial support and political alignment of physicists in Weimar Germany', *Minerva*, 12 (1), 1974, 39–66.

Forman, Paul and Sánchez-Ron, José M. (eds.), *National military establishments and the advancement of science and technology: studies in 20th century history*, Dordrecht; Boston: Kluwer Academic Publishers, 1996.

Fortescue, Stephen, *Science policy in the Soviet Union*, London: Routledge, 1990.

Fox, Robert, 'Scientific enterprise and the patronage of research in France 1800–70', *Minerva*, 11 (4), 1973, 442–73.

'Science and government' in Porter (ed.), *Eighteenth-century science*, pp. 107–28.

The savant and the state: science and cultural politics in nineteenth-century France, Baltimore: Johns Hopkins University Press, 2012.

Fox, Robert and Weisz, George (ed.), *The organization of science and technology in France, 1808–1914*, Cambridge: Cambridge University Press, 1980.

Frängsmyr, Tore (ed.), *Solomon's House revisited: the organisation and institutiona-lisation of science*, Canton: Science History Publications, 1990.

Frängsmyr, Tore, Heilbron, J. L. and Rider, Robin E. (eds.), *The quantifying spirit in the eighteenth century*, Berkeley: University of California Press, 1990.

Frates da Costa, Palmira and Leitão, Henrique, 'Portuguese imperial science, 1450–1800: A historiographical review', in Daniela Bleichmar, Paul De Vos, Kristin Huffine and Kevin Sheehan (eds.), *Science*, pp. 35–53.

Frieman, Wendy, 'People's Republic of China: between autarky and interdepen-dence' in Solingen (ed.), *Scientists and the state*, pp. 127–144.

Galbreath, Ross, *DSIR. Making science work for New Zealand*, Wellington: Victoria University Press, 1998.

Gascoigne, John, 'A reappraisal of the role of the universities in the Scientific Revolution' in D. C. Lindberg and R. S. Westman (eds.), *Reappraisals of the Scientific Revolution*, Cambridge: Cambridge University Press, 1990, pp. 207–60.

Science in the service of empire: Joseph Banks, the British state and the uses of science in the age of revolution, Cambridge: Cambridge University Press, 1998.

'Joseph Banks, mapping and the geographies of natural knowledge' in Miles Ogden and Charles Withers (eds.), *Georgian geographies: Essays on space, place and landscape in the eighteenth century*, Manchester: Manchester University Press, 2004, pp. 151–73.

'The Royal Society, natural history and the peoples of the "New World(s)", 1660–1800', *British Journal for the History of Science*, 42 (4), 2009. 539–62.

'Crossing the pillars of Hercules: Francis Bacon, the scientific revolution and the New World' in Ofer Gal and Raz Chen-Morris (eds.), *Baroque Science*, Dordrecht: Springer, 2013, pp. 217–38.

'Science and the British empire from its beginnings to 1850' in Bennett and Hodge (eds.), *Science and empire*, pp. 47–67.

Encountering the Pacific in the age of the Enlightenment, Cambridge: Cambridge University Press, 2014.

'From science to religion: justifying French Pacific voyaging and expansion in the period of the Restoration and the July Monarchy', *Journal of Pacific History*, 50, 2015, 109–27.

Gaukroger, Stephen, *Francis Bacon and the transformation of early-modern philoso-phy*, Cambridge: Cambridge University Press, 2001.

The emergence of a scientific culture: Science and the shaping of modernity, 1210–1685, Oxford: Clarendon Press, 2006.

Genuth, Joel, 'Groping towards science policy in the United States in the 1930s', *Minerva*, 25 (3), 1987, 238–68.

Geyer-Kordesch, Johanna, 'Court physicians and state regulation in eighteenth-century Prussia: the emergence of medical science and the demystification of the body' in Nutton (ed.), *Medicine at the courts*, pp. 1–48.

Gillispie, Charles, 'Science and politics, with special reference to Revolutionary and Napoleonic France', *History of Technology*, 4, 1987, 213–23.

Science and polity in France: The Revolutionary and Napoleonic years, Princeton: Princeton University Press, 2004.

Gilmartin, David, 'Scientific empire and imperial science: colonialism and irrigation technology in the Indus basin', *The journal of Asian studies*, 53 (4), 1994, 1127–49.

Gilpin, Robert, *France in the age of the scientific state*, Princeton: Princeton University Press, 1968.

Glass, D. V., 'John Graunt and his natural and political observations', *Notes and Records of the Royal Society of London*, 19 (1), 1964, 63–100.

Gleason, Mary, *The Royal Society of London, years of reform 1827–1847*, New York: Garland, 1991.

Godlewska, Anne, *Geography unbound: French geographic science from Cassini to Humboldt*, Chicago: Chicago University Press, 1999.

Goodman, David, 'Philip II's patronage of science and engineering', *The British journal for the history of science*, 16 (1), 1983, 49–66.

  Power and penury: government, technology, and science in Philip II's Spain, Cambridge: Cambridge University Press, 1988.

  'Science, medicine and technology in Colonial Spanish America. New interpretations, new approaches' in Bleichmar, De Vos, Huffine, Kristin and Sheehan (eds.), *Science in the Spanish and Portuguese empires*, pp. 9–34.

Goodman, David and Russell, Colin A., *The rise of scientific Europe, 1500–1800*, Milton Keynes: Open University, 1991.

Gordin, Michael D., 'The importation of being earnest: The early St. Petersburg Academy of Sciences', *Isis*, 91 (1), 2000, pp. 1–31.

Graham, Loren, *The Soviet Academy of Sciences and the Communist Party, 1927–1932*, Princeton: Princeton University Press, 1967.

  'Development of science policy in the Soviet Union' in Dixon T. Long and Christopher Wright,(eds.), *Science policies of industrial nations*, New York: Praeger, 1975, pp.12–58.

  Science, philosophy, and human behaviour in the Soviet Union, New York: Columbia University Press, 1987.

  Science in Russia and the Soviet Union: A short history, Cambridge: Cambridge University Press, 1993.

  What have we learned about science and technology from the Russian experience?, Stanford: Stanford University Press, 1998.

Graham, Loren and Dezhina, Irina, *Science in the new Russia: Crisis, aid, reform*, Bloomington: Indiana University Press, 2008.

Gray, George W., *Science at war*, Freeport: Books for Libraries, 1972.

Greenberg, Daniel S., *The politics of pure science*, New York: 1969.

Grove, J. W., *In defence of science: science, technology, and politics in modern society*, University of Toronto Press, 1989.

Grunden, Walter E., Kawamura, Yutaka, Kolchinsky, Eduard, Maier, Helmut and Yamazaki, Masakatsu, 'Laying the foundation for wartime research: a comparative overview of science mobilization in National Socialist Germany, Japan, and the Soviet Union', *Osiris*, 20, 2005, 79–106.

Guerlac, Henry E., 'Science and French national strength' in E. Earle (ed.), *Modern France*, Princeton: Princeton University Press, pp. 81–105.

Gummett, Philip, *Scientists in Whitehall*, Manchester: Manchester University Press, 1980.

Gustafson, Thane, 'Why doesn't Soviet science do better than it does?' In Linds
    L. Lubrao and Susan Gross Solomon (eds.), *The social context of Soviet science*,
    Boulder: Westview, 1980, pp. 31–67.
Haber, L. F., 'Government intervention at the frontiers of science: British
    dyestuffs and synthetic organic chemicals 1914–39', *Minerva*, 11 (1),
    1973, 79–94.
Hahn, Roger, *The anatomy of a scientific institution: the Paris Academy of Sciences,
    1666–1803*, Berkeley: University of California Press, 1971.
    'The age of academies' in Frängsmyr (ed.), *Solomon's House*, pp. 3–12.
    'Louis XIV's science policy' in David Rubin (ed.), *Sun King: The ascendancy of
    French culture during the reign of Louis XIV*, Washington, DC: Folger
    Shakespeare Library, 1992, pp. 195–207.
Hall, A. R., 'Gunnery, science and the Royal Society' in John G. Burke (ed.),
    The uses of science in the age of Newton, Berkeley, 1983, pp. 111–42.
Hall, M. B., *All scientists now: The Royal Society in the nineteenth century*,
    Cambridge: Cambridge University Press, 1984.
Hamblin, Jacob Darwin, 'Visions of international scientific cooperation: the case
    of oceanic science', *Minerva*, 38, 2000, 393–423.
Harrison, Carol E., 'Projections of the Revolutionary nation: French expeditions
    in the Pacific, 1791–1803', *Osiris*, 24 (1), 2009, 33–52.
Harrison, Carol E. and Johnson, Ann, 'Introduction: science and national
    identity', *Osiris*, 24 (1), 2009, pp. 1–14.
Hartcup, Guy, *The war of invention. Scientific developments, 1914–18*, London:
    Brassy, 1988.
    Effect of science on the Second World War, Houndmills: Palgrave, 2000.
Harwood, Jonathan, 'Universities' in Bowler and Pickstone (eds.), *The Cambridge
    history of science*, vol. 6, pp. 90–107.
Hayton, Darin, *The crown and the cosmos: Astrology and the politics of Maximilian I*,
    Pittsburgh: University of Pittsburgh Press, 2015.
Headrick, Daniel, *Tools of empire: Technology and European imperialism in the
    nineteenth century*, Oxford and New York: Oxford University Press, 1981.
    The tentacles of progress. Technology transfer in the age of imperialism, 1850–1940,
    Oxford: Oxford University Press, 1988.
Heilbron, John L., 'The measure of enlightenment' in Frängsmyr, Heilbron and
    Rider (eds.), The quantifying spirit, pp. 207–19.
Heim, Susanne, Sachse, Carola and Walker, Mark (eds.), *The Kaiser Wilhelm
    Society under National Socialism*, Cambridge: Cambridge University Press,
    2009.
Hevly, Bruce, 'The tools of science: radio, rockets, and the science of naval
    warfare' in Forman and Sánchez-Ron (eds.), *National military establishments*,
    pp. 215–32.
Higgs, Edward, 'The struggle for the occupational census, 1841–1911' in
    Roy MacLeod (ed.), *Government and expertise: Specialists, administrators and
    professionals, 1860–1919*, Cambridge: Cambridge University Press, 1988, pp.
    73–86.
Higueras, Maria Dolores, *The northwest coast of America: iconographic album of the
    Malaspina expedition*, Madrid: Muso Naval, 1991.

Hilaire-Pérez, Liliane, 'Invention and the state in 18th-Century France', *Technology and culture*, 32 (4), 1991, 911–31.

Hobbes, Thomas, *Leviathan*, ed., C. B. Macpherson, Harmondsworth: Penguin, 1968.

Hodge, Joseph M., 'Science and empire: an overview of the historical scholarship' in Bennett and Hodge (eds.), *Science and empire*, pp. 3–29.

Hoffmann, Dieter, 'Between autonomy and accommodation: the German Physical Society during the Third Reich', *Physics in Perspective*, 7, 2005, 293–329.

Holloway, David, 'Entering the arms race: the Soviet decision to build the atomic bomb, 1939–45', *Social Studies of Science*, 11 (2), 1981, 159–97.
  Stalin and the bomb: The Soviet Union and atomic energy 1939–1956, New Haven: Yale University Press, 1994.

Hughes, Arnold, 'The nation-state in Black Africa' in Tivey (ed.), *The nation-state*, pp. 122–47.

Hunter, Michael, *Science and society in Restoration science*, Cambridge: Cambridge University Press, 1981.
  'First steps in institutionalization: the role of the Royal Society of London' in Frängsmyr (ed.), *Solomon's House*, pp. 13–30.

Hutchinson, Eric, 'Scientists and civil servants: the struggle over the National Physical Laboratory in 1918', *Minerva*, 7 (3), 1969, 373–98.
  'Scientists as an inferior class: the early years of the DSIR', *Minerva*, 8 (3), 1970, 396–411.
  'Government laboratories and the influence of organized scientists', *Science Studies*, 1, 1971, 331–56.
  'Origins of the University Grants Committee', *Minerva*, 1975, 13 (4), 583–620.

Jami, Catherine, 'Western mathematics in China. Seventeenth century and nineteenth century' in Petitjean, Jami and Moulin (eds.) *Science and empires*, 79–88.

Johannisson, Karin, 'Society in numbers: the debate over quantification in eighteenth century political economy' in Frängsmyr, Heilbron and Rider (eds.), *The quantifying spirit*, pp. 343–61.

Johnson, Jeffrey A., 'Academic chemistry in Imperial Germany', *Isis*, 76 (4), 1985, 500–24.
  'Technological mobilization and munitions production. Comparative perspectives on Germany and Austria' in Roy MacLeod and Jeffrey A. Johnson (eds.), *Frontline and factory*, pp. 1–20.

Josephson, Paul R., *Physics and politics in revolutionary Russia*, Berkeley: University of California Press, 1991.
  'The political economy of Soviet science from Lenin to Gorbachev' in Solingen (ed.), *Scientists and the state*, pp. 145–169.
  Totalitarian science and technology, Humanities Press, Atlantic Highlands, NJ, 1996.
  'Science, ideology, and the state physics in the twentieth century' in Mary Jo Nye (ed.), *Cambridge history of science*, vol. 5, pp. 579–97.

Kapila, Shruti, 'The enchantment of science in India', *Isis*, 101 (1), 2010, 120–32.

Kapur, Ashok, 'India: the nuclear scientists and the state, the Nehru and post-Nehru years', in Solingen (ed.), *Scientists and the state*, pp. 209–29.

Kenney-Wallace, Geraldine A. and Mustard, J. Fraser, 'From paradox to paradigm: the evolution of science and technology in Canada', *Daedalus*, 117 (4), 1988, 191–214.

Kevles, Daniel J., 'The National Science Foundation and the debate over postwar research policy, 1942–1945: a political interpretation of *Science – the endless frontier*', *Isis*, 68 (1), 1977, 4–26.

The physicists. The history of a scientific community in Modern America, New York: Alfred A. Knopf, 1978.

Kiernan, Michael (ed.), *The Oxford Francis Bacon*, IV: *The advancement of learning* Oxford: Clarendon Press, 2000.

King, James, *Science and rationalism in the government of Louis XIV*, New York: Octagon Books, 1972 [1949].

Kirby, William C., 'Technocratic organisation and technological development in China: The Nationalist experience and legacy, 1928–1953' in Denis Simon and Merle Goldman (eds.), *Science and technology in Post-Mao China*, Cambridge, MA: Harvard University Press, 1988, pp. 23–44.

Knight, David, *The nature of science: the history of science in western culture since 1600*, London: A. Deutsch, 1976.

Kohler, Robert, *Partners in science: Foundations and national scientists 1900–1945*, Chicago: Chicago University Press, 1991.

Kojevnikov, Alexei, 'The Great War, the Russian Civil War, and the invention of Big Science', *Science in Context*, 15 (2), 2002, pp. 239–75.

Krementsov, Nickolai, *Stalinist science*, Princeton: Princeton University Press, 1997.

'Russian science in the twentieth century' in John Krige and Dominique Pestre (eds.), *Science in the twentieth century*, pp. 777–94.

Krige, John and Pestre, Dominique (eds.), *Science in the twentieth century*, Amsterdam: Harwood Academic Publishers, 1997.

Krishna, V. V., 'The colonial "model" and the emergence of national science in India: 1876–1920', in Petitjean, Jami and Moulin, (eds.), *Science and empires*, pp. 57–72.

Kumar, Deepak (ed.), *Science and empire: Essays in Indian context (1700–1947)*, Delhi: Anamika Prakashan, 1991.

Science and the Raj, 1857–1905, Delhi: Oxford University Press, 1995.

Kümmel, Werner, '*De morbis aulicis*; on diseases found at court' in Nutton (ed.), *Medicine and the courts*, pp. 1–48.

Lacey, Michael, 'The world of the bureaus: government and the positivist project in the late nineteenth century' in Michael Lacey and Mary Furner (eds.), *The state and social investigation in Britain and the United States*, Cambridge: Cambridge University Press, 1993, pp. 127–70.

Lafuente, Antonio, 'Enlightenment in an imperial context; local science in the late- eighteenth-century Hispanic world, *Osiris* 2nd ser., 15, 2000, 155–73.

Laird, W. R., 'Patronage of mechanics and theories of impact in sixteenth-century Italy' in Moran, *Patronage and institutions*, pp. 51–66.

Lakoff, Sanford, *Knowledge and power: essays on science and government*, New York: The Free Press, 1966.

'The scientific establishment and American pluralism', in Lakoff, *Knowledge and power*, pp. 377–394.

Langines, Janis, *Conserving the Enlightenment. French military engineering from Vauban to the Revolution*, Cambridge, MA: MIT Press, 2004.

Lapp, Ralph E., *The new priesthood: The scientific elite and the uses of power*, New York: Harper & Row, 1965.

Largent, Mark A, *Breeding contempt: The history of coerced sterilization in the United States*, New Brunswick: Rutgers University Press, 2008.

Lassman, Thomas C., 'Government science in postwar America: Henry A. Wallace, Edward U. Condon, and the transformation of the National Bureau of Standards, 1945–1951', *Isis*, 96 (1), 2005, 25–51.

Leary, John E., *Francis Bacon and the politics of science*, Ames: Iowa State University Press, 1994.

Lehmann, Jörg, Morselli, Francesca, *Science and technology in the First World War*. CENDARI Archival Research Guide. 2016.

Le Roux, Thomas, 'Chemistry and industrial and environmental governance in France 1770–1830', *History of Science*, 54 (2), 2016, 195–222.

Levere, Trevor H. and Jarrell, Richard A., (eds.), *A curious field-book: science & society in Canadian history*, Toronto: Oxford University Press, 1974.

Lewis, Robert, *Science and industrialisation in the USSR*, New York, 1979.

Limoges, Camille, 'The development of the Museum d'histoire naturelle of Paris, c. 1800–1914' in Fox and Weisz (eds.), *The organization of science*, pp. 213–40.

Long, T. Dixon, 'Policy and politics in Japanese science: the persistence of a tradition', *Minerva*, 7 (3), 1969, 426–53.

Low, Morris, Nakayama, Shigeru and Yoshioka, Hitoshi, *Science, technology and society in contemporary Japan*, Cambridge: Cambridge University Press, 1999.

Lowood, Henry, 'The calculating forester: quantification, cameral science, and the emergence of scientific forestry management in Germany' in Frängsmyr, Heilbron and Rider (eds.), *The quantifying spirit*, pp. 315–42.

Lux, David, 'The reorganisation of science 1450–1700' in Moran (ed.), *Patronage*, pp. 185–94.

Lyton, Edwin, 'Mirror-image twins: the communities of science and technology' in Daniels (ed.), *Nineteenth-century American science*, pp. 210–230.

MacDonald, Lee T., 'Making Kew Observatory: the Royal Society, the British Association and the politics of early Victorian science', *British Journal of the History of Science*, 48 (3), 2015, 409–33.

MacLeod, Roy, 'The Alkali Acts administration, 1863–4: the emergence of the civil scientist', *Victorian Studies*, 9 (2), 1965, 85–112.

'Of medals and men: a reward system in Victorian science, 1826–1914', *Notes and Records of the Royal Society of London*, 26, 1971, 81–105.

'The Royal Society and the Government Grant: notes on the administration of scientific research, 1849–1914', *The Historical Journal*, 14 (2), 1971, 323–58.

'Resources of science in Victorian England: the Endowment of Science Movement, 1868–1900' in D. Mathias (ed.), *Science and society 1600–1900*, Cambridge: Cambridge University Press, 1972, pp. 111–66.

'Scientific advice for British India: imperial perceptions and administrative goals, 1898–1923', *Modern Asian Studies*, 9 (3), 1975, 343–84.

'Science and the Treasury: principles, personalities and politics, 1870–85' in G. L. E. Turner (ed.), *The patronage of science in the nineteenth century*, Leyden: Noordhoff, 1976, pp. 115–72.

'Whigs and savants: reflections on the reform movement in the Royal Society, 1830–1848' in Ian Inkster and John Morrell (eds.), *Metropolis and province: science in British culture 1780–1850*, London, 1983, pp. 55–90.

'Imperial reflections in the Southern Seas: the Funafuti expedition 1896–1904' in Roy MacLeod and Philip F. Rehbock (eds.), *Nature in its greatest extent. Western science in the Pacific*, Honolulu: University of Hawaii Press, 1988, pp. 159–91.

'The "Arsenal" in the strand: Australian chemists and the British munitions effort 1916–1919', *Annals of Science*, 46 (1), 1989, 45–67.

'"Combat scientists": the Office of Scientific Research and Development and Field Service in the Pacific', *War & Society*, 11 (2), 1993, 117–34.

'Discovery and exploration' in Bowler and Pickstone (eds.), *The Cambridge history of science*, vol. 6, pp. 34–59.

'The scientists go to war: revisiting precept and practice, 1914–1919', *Journal of War & Culture Studies*, 2 (1), 2009, 37–51.

MacLeod, Roy and Andrews, E. Kay, 'The origins of the D.S.I.R.: reflections on ideas and men, 1915–1916', *Public Administration*, 48 (1), 1970, 23–48.

'Scientific advice in the War at Sea, 1915–1917: the Board of Invention and Research', *Journal of Contemporary History*, 6 (2), 1971, 3–40.

MacLeod, Roy and Johnson, Jeffrey A. (eds.), *Frontline and factory: Comparative perspectives on the chemical industry at war 1914–1924*, Dordrecht: Springer, 2006.

'Introduction' in MacLeod and Johnson (eds.), *Frontline and factory*, pp. xiii–xix.

MacLeod, Roy and MacLeod, Kay, 'War and economic development: governments and the optical industry in Britain, 1914–18' in J. M. Winter (ed.), *War and economic development: Essays in memory of David Joslin*, Cambridge: Cambridge University Press, 1975, pp. 165–203.

Macpherson, C. B., 'Introduction' to Thomas Hobbes, *Leviathan*, pp. 9–63.

Macrakis, Kristie, *Surviving the swastika: Scientific research in Nazi Germany*, New York: Oxford University Press, 1993.

Maerker, Anna, 'Political order and the ambivalence of expertise: Count Rumford and welfare reform in late eighteenth-century Munich', *Osiris*, 25 (1), 2010, 213–30.

Makarova, Raisa (trans. and ed. by R. A. Pierce and A. S. Donnelly), *Russians in the Pacific 1743–1799*, Kingston: Limestone Press, 1975.

Manning, Thomas, *US Coast Survey vs Naval Hydrographic Office: A nineteenth-century century rivalry in science and politics*, Tuscaloosa: University of Alabama Press, 1988.

Martin, Julian, *Francis Bacon, the state, and the reform of natural philosophy*, Cambridge: Cambridge University Press, 1992.

McClellan, James, *Science reorganised. Scientific societies in the eighteenth century*, New York: Columbia University Press, 1985.

    'Scientific institutions and the organization of science' in Roy Porter (ed.), *Eighteenth- Century science*, pp. 86–106.

    Colonialism and science: Saint Domingue in the old regime, Chicago: University of Chicago, 2010.

McClellan, James and Regourd, François, *The colonial machine: French science and overseas expansion in the old regime*, Ternhout, Belgium: Brepolis, 2011.

McClelland, Charles, *State, society, and university in Germany 1700–1914*, Cambridge: Cambridge University Press, 1980.

McGrath, Patrick J., *Scientists, business, and the state 1890–1960*, Chapel Hill: University of North Carolina Press, 2002.

McGucken, William, 'Freedom and planning, on freedom and planning in science: the Society for Freedom in Science, 1940–46', *Minerva*, 1978, 16 (1), 42–72.

McGucken, William, 'Royal Society and the Scientific Advisory Committee to Britain's War Cabinet, 1939–1940', *Notes and Records of the Royal Society*, 33 (1), 1978, 87–115.

    'Central organisation of scientific and technical advice in the United Kingdom during the Second World War', *Minerva*, 13, 1979, 33–69.

Mehrtens, Herbert, 'The social system of mathematics and National Socialism: a survey' in Renneberg and Walker (eds.), *Science, technology and National Socialism*, pp. 291–311.

Mendelsohn, Everett, Smith, Merritt Roe and Weingart, Peter (eds.), *Science, technology and the military*, Dordrecht: Springer, 1988.

Meyer-Thurow, Georg, 'The industrialization of invention: a case study from the German chemical industry', *Isis*, 73, 1982, 363–81.

Miller, Howard S., 'The political economy of science' in Daniels (ed.), *Nineteenth-century American science*, pp. 95–112.

Miller, H. Lyman, *Science and dissent in post-Mao China: The politics of knowledge*, Seattle: University of Washington, 1996.

Moore, Kelly, *Disrupting science. Social movements, American scientists, and the politics of the military, 1945–1975*, Princeton: Princeton University Press, 2008.

Moran, Bruce T., *Patronage and institutions: science, technology, and medicine at the European court, 1500–1750*, Rochester, New York: Boydell Press, 1991, pp. 23–50.

    'Patronage and institutions: courts, universities and academies in Germany; an overview' in Moran, *Patronage and institutions*, pp. 169–83.

    The alchemical world of the German Court: Occult philosophy and chemical medicine in the circle of Moritz of Hessen, Stuttgart: Franz Steiner, 1991.

    'Courts and academies' in Park and Daston (eds.), *The Cambridge history of science*, vol. 3, pp. 251–71.

Moseley, Russell, 'The origins and early years of the National Physical Laboratory: a chapter in the pre-history of British science policy', *Minerva*, 16, 1978, 222–50.

Mukerji, Chandra, *A fragile power: Scientists and the state*, Princeton University Press, 1989.

Nakayama, Shigeru, *Science, technology and society in postwar Japan*, London: Kegan Paul, 1991.

Navari, Cornelia, 'The origins of the nation-state' in Tivey, *The nation-state*, pp. 13–38.

Naylor, Simon, 'Log books and the law of storms: maritime meteorology and the British Admiralty in the nineteenth century', *Isis* 106 (4), 2015, 771–97.

Nelson, Brian, *The making of the modern state: a theoretical evaluation*, London: Palgrave, 2006.

Neufeld, Michael J., 'The guided missile and the Third Reich: Peenemünde and the forging of a technological revolution' in Renneberg and Walker (eds.), *Science, technology and national socialism*, pp. 51–71.

Nutton, Vivian (ed.), *Medicine and the courts of Europe, 1500–1837*, London: Routledge, 1990.

Nye, Mary Jo (ed.), *The Cambridge history of science*, vol. 5, *The modern physical and mathematical sciences*, Cambridge: Cambridge University Press, 2002.

Olby, R. C., Cantor, G. N., Christie, J. R. R. and Hodge, M. J. S. (eds.), *Companion to the history of modern science*, London: Routledge, 1990.

Ortiz, Eduardo L., 'Army and science in Argentina: 1850–1950' in Forman and Sánchez-Ron (eds.), *National military establishments*, pp. 153–84.

Osborne, Michael A., *Nature, the exotic, and the science of French colonialism*, Bloomington: Indiana University Press, 1994.

Outram, Dorinda, 'Politics and vocation. French science, 1793–1830, *British Journal for the History of Science*, 13 (1), 1980, 27–43.

Owens, Larry, 'Science in the United States' in Krige and Pestre (eds.), *Science in the twentieth century*, pp. 821–37.

Park, Katharine and Daston, Lorraine (eds.), *The Cambridge history of science*, vol. 3, Cambridge: Cambridge University Press, 2006.

Pattison, Michael, 'Scientists, inventors and the military in Britain, 1915–19: The Munitions Inventions Department', *Social Studies of Science*, 13 (4), 1983, 521–68.

Paul, Harry, *From knowledge to power: the rise of the science empire in France, 1860–1939*, Cambridge: Cambridge University Press, 1985.

Penrick, James L., Pursell, Carroll W., Sherwood, Margaret and Swain, Donald (eds.), *The politics of American science*, Cambridge, MA: MIT Press, 1972.

Petitjean, Patrick, 'Science and the "Civilizing Mission": France and the colonial enterprise' in Benedikt Stutchtey (ed.), *Science across the European Empires – 1800–1950*, Oxford University Press, 2005. pp. 107–28,

Petitjean, Patrick, Jami, Catherine and Moulin, A. M. (eds.), *Science and empires: Historical studies about scientific development and European expansion*, Dordrecht: Springer, 1992.

Pfetsch, Frank, 'Scientific organisation and science policy in Imperial Germany, 1871–1914: the foundation of the Imperial Institute of Physics and Technology', *Minerva*, 8 (4), 1970, 557–80.

'Germany: three models of interaction – Weimar, Nazi, Federal Republic' in Solingen (ed.), *Scientists and the state*, pp. 189–208.

Pielke, Roger, '*In retrospect: Science – the endless frontier*', *Nature*, 466, 2010, 922–23.

Pierre, Andrew J., *Nuclear politics, the British experience with an independent nuclear force*, London: Oxford University Press, 1972.

Pimentel, Juan, 'The Iberian vision: science and empire in the framework of a universal monarchy, 1500–1800', *Osiris*, 15, 17–30.

Poggi, Gianfranco, 'The modern state and the idea of progress' in Gabriel Almond, Martin Chodorow and Roy Harvey Pearce (eds.), *Progress and its discontents*, Berkeley: University of California Press, 1982, pp. 337–60.

Poletschek, Silvia, 'The invention of Humboldt and the impact of National Socialism: the German university idea in the first half of the twentieth century' in Margit Szöllösi-Janze (ed.), *Science in the Third Reich*, Oxford: Berg, 2001, pp. 27–58.

Pollock, Ethan, *Stalin and the Soviet science wars*, Princeton: Princeton University Press, 2006.

Porter, Dorothy, *Health, civilization, and the state: a history of public health from ancient to modern times*, London: Routledge 1999.

Porter, Roy, 'The scientific revolution: a spoke in the wheel?' in Roy Porter and Mikuláš Teich (eds.), *Revolutions in history*, Cambridge: Cambridge University Press, 1980, pp. 290–316.

Porter, Roy (ed.), *Cambridge history of science*, vol. *4. Eighteenth-century science*, Cambridge: Cambridge University Press, 2003.

Porter, Roy and Teich, Mikuláš (eds.), *The scientific revolution in national context*, Cambridge: Cambridge University Press, 1992.

Porter, Theodore M., *The rise of statistical thinking 1820–1900*, Princeton University Press, 1986.

Trust in numbers. The pursuit of objectivity in science and public life, Princeton: Princeton University Press, 1995.

Prakash, Gyan, *Another reason. Science and the imagination of modern India*, Princeton: Princeton University Press, 1999.

Price, Don K., 'The scientific establishment' in Robert Gilpin and Christopher Wright (eds.), *Scientists and national policy-making*, Columbia University Press, New York, 1964. pp. 19–40.

The scientific estate, Cambridge, MA: Harvard University Press, 1965.

Pursell, Carroll W. Jr., 'The anatomy of a failure: the Science Advisory Board, 1933–1935', *Proceedings of the American Philosophical Society*, 109 (6), 1965, 342–51.

'A preface to government support of research and development: research legislation and the National Bureau of Standards, 1935–41', *Technology and Culture*, 9 (2), 1968, pp. 145–64.

'Science agencies in World War II: the OSRD and its challengers' in Nathan Reingold (ed.), *The sciences in the American context*, pp. 359–78.

Pyatt, Edward, *The National Physical Laboratory. A history*, Bristol: Adam Hilger, 1983.

Pyenson, Lewis, 'On the military and the exact sciences in France' in Forman and Sánchez-Ron (eds.), *National military establishments*, pp. 135–52.

Pyenson, Lewis and Sheets-Pyenson, Susan, *Servants of nature: a history of scientific institutions, enterprises, and sensibilities*, London: Fontana, 1999.

Rabb, Theodore K., *The struggle for stability in early modern Europe*, New York: Oxford University Press 1975.

Raj, Kapil, 'Knowledge, power and modern science: the Brahmins strike back' in Kumar (ed.) *Science and empire*, pp. 115–25.

Rappaport, Rhoda, 'The liberties of the Paris Academy of Sciences, 1716–1785' in Harry Woolf (ed.), *The Analytic spirit. Essays in the history of science in honor of Henry Guerlac*, Ithaca: Cornell University Press, 1981, pp. 225–56.

Rasmussen, Anne, 'Science and technology' in John Horne (ed.), *A companion to World War I*, Chichester: Wiley Blackwell, pp. 307–22.

Rees, Graham (ed.), *The Oxford Francis Bacon. The instauratio magna. Part II: Novum organum and associated texts*, Oxford: Clarendon, 2000.

Reich, Leonard S., *The making of American industrial research: Science and business at GE and Bell, 1876–1926*, Cambridge: Cambridge University Press, 1985.

Reidy, Michael S., *Tides of history: Ocean science and Her Majesty's navy*, Chicago: University of Chicago Press, 2008.

Reidy, Michael S., Kroll, Gary and Conway, Erik M., *Exploration and science: social impact and interaction*, Santa Barbara: ABC-CLIO, 2007.

Reingold, Nathan (ed.), *The sciences in the American context: new perspectives*, Washington, DC: Smithsonian Institution Press, 1979.

Renneberg, Monika and Walker, Mark (eds.), *Science, technology and national socialism*, Cambridge: Cambridge University Press, 1994.

Resnik, David, *Playing politics with science: Balancing scientific independence and government oversight*, Oxford University Press, 2009.

Richards, Pamela, 'Great Britain and Allied scientific information, 1939–45', *Minerva*, 26 (2), 1988, 177–98.

Roberts, Lissa, 'Le centre de toutes choses: Constructing and managing centralization on the Isle de France', *History of Science*, 52 (3), 2014, 319–42.

Roland, Alex, 'Science and war', Osiris, 1, 1985, 247–72.
'Science, technology and war' in Nye (ed.), *Cambridge history of science*, vol. 5, pp. 559–78.

Ronayne, Jarleth, *Science in government*, London: Edward Arnold, 1984.

Rose, Hilary and Rose, Steven, *Science and society*, London: Allen Lane, 1969.

Rosenberg, Charles E., 'Science, technology, and economic growth: the case of the agricultural experiment station scientists, 1875–1914' in George Daniels (ed.), *Nineteenth-century American science*, pp. 181–209.

Rusnock, Andrea, 'Quantification, precision, and accuracy: determinations of population in the ancient regime' in M. Norton Wise (ed.), *The values of precision*, Princeton: Princeton University Press, 1995, pp. 17–38.

'Biopolitics: political arithmetic in the Enlightenment' in William Clark, Jan Golinsi and Simon Schaffer (eds.), *The sciences in enlightened Europe*, Chicago: Chicago University Press, 1999, pp. 50–68.

Russell, Colin, *Science and social change in Britain and Europe, 1700–1900*, London: Macmillan, 1984.

Saich, Tony, 'Reform of China's science and technology organisational system' in Simon and Goldman (eds.), *Science and technology*, pp. 69–88.

Salomon, Jean-Jacques, *Science and politics*, Cambridge, MA: Harvard University Press, 1973.

Sapolsky, Harvey M., *Science and the navy: The history of the Office of Naval Research*, Princeton: Princeton University Press, 1990.

Sargent, Rose-Mary, 'Bacon as an advocate for cooperative scientific research' in Markku Peltonen (ed.), *The Cambridge companion to Bacon*, Cambridge: Cambridge University Press, 1996, pp. 147–71.

Schaffer, Simon, 'What is science?' in Krige and Pestre (eds.), *Science in the twentieth century*, pp. 27–41.

Schedvin, Carl B., *Shaping science and industry: a history of Australia's Council for Scientific and Industrial Research, 1926–49*, Sydney: Allen & Unwin, 1987.

Schiavon, Martina, 'The Bureau des longitudes: an institutional study' in Rebekah Higgitt, Richard Dunn and Peter Jones (eds.), *Navigational enterprises in Europe and its empires, 1730–1850*, Basingstoke: Palgrave, pp. 65–85.

Schiebinger, Londa, 'Women of natural knowledge' in Park and Daston (eds.), Cambridge history of science, vol. 3, pp. 192–205.

Schiebinger, Londa and Swan, Claudia, 'Introduction' in Londa Schiebinger and Claudia Swan (eds.), *Colonial botany. Science, commerce and politics in the early modern world*, Philadelphia: University of Pennsylvania Press, 2005, pp. 1–16.

Schneider, L. A., 'Learning from Russia: Lysenkoism and the fate of genetics in China, 1950–1986' in D. Simon and M. Goldman (eds.), *Science and technology*, pp. 45–65.

Schoeder-Gudehus, Brigitte, 'The argument for the self-government and public support of science in Weimar Germany, *Minerva*, 10 (4), 1972, 537–70.

'Nationalism and internationalism' in Olby, Cantor, Christie and Hodge (eds.), *Companion*, pp. 909–19.

Schwartzman, Simon, 'Brazil: scientists and the state—evolving models and the 'great leap forward' in Solingen (ed.), *Scientists and the state*, pp. 171–88.

Schweber, S., 'The mutual embrace of science and the military: ONR and the growth of physics in the United States after World War II' in E. Mendelsohn, Merritt Roe Smith and Peter Weingart (eds.), *Science, technology and the military*, II, pp. 3–45.

Scott, James, *Seeing like a state: How certain schemes to improve the human condition have failed*, New Haven: Yale University Press, 1998.

Scott-Carver, J., 'A reconsideration of eighteenth-century Russia's contributions to European science' in Malcolm Osler (ed.), *Science in Europe, 1500–1800:*

a secondary sources reader, Houndmills, Basingstoke: Palgrave, 2002, pp. 226–33.

Seidel, Robert, 'The origins of the Lawrence Berkeley Laboratory' in Peter Galison and Bruce Hevly (ed.), *Big science: The growth of large-scale research*, Stanford: Stanford University Press, 1992, pp. 21–45.

Shapin, Stephen, *The scientific revolution*, Chicago: Chicago University Press, 1996.

Shapin, Steven and Schaffer, Simon, *Leviathan and the air-pump: Hobbes, Boyle, and the experimental life*, Princeton: Princeton University Press, 1985.

Shakelford, Jole, 'Paracelsianism and patronage in early modern Denmark' in Moran (ed.), Patronage and institutions, pp. 85–109.

Shils, Edward, 'The order of learning in the United States from 1865 to 1920: the ascendancy of the universities', *Minerva*, 16 (2), 1978, pp. 159–95.

Shinn, Terry, 'The industry, research, and education nexus' in Nye (ed.), *Cambridge history of science*, vol. 5, 133–53.

Siegelbaum, Lewis H. *The politics of industrial mobilization in Russia, 1914–17: a study of the war-industries committees*, Houndmills: Palgrave, 1983.

Simon, D. and Goldman, M. (eds.), *Science and technology in post-Mao China*, Cambridge, MA: Harvard University Press, 1989.

'The onset' in Simon and Goldman (eds.), *Science and Technology*, pp. 3–20.

Sinclair, Bruce, 'The promise of the future: technical education' in George Daniels (ed.), *Nineteenth-century American science: A reappraisal*, Evanston: Northwestern University Press, 1972, pp. 249–72.

Sinha, Jagdish N., *Science, war and imperialism. India in the Second World War*, Leiden: Brill, 2008.

Sivasundaram, Sujit, 'Sciences and the global: On methods, questions, and theory', *Isis*, 101 (1), 2010, 146–58.

Skinner, Quentin, 'The state' in T. Ball, T. Farr and R. Hanson (eds.), *Political innovation and conceptual change*, Cambridge: Cambridge University Press, 1989, pp. 90–131.

Smith, Bruce L. R., *American science policy since World War II*, Washington, DC: Brookings Institution, 1990.

'The United States: the formation and breakdown of the postwar government-science compact' in Solingen (ed.), *Scientists and the state*, pp. 33–61.

Smith, John Denly, ' World War II and the transformation of the American chemical industry' in E. Mendelsohn, Merritt Roe Smith and Peter Weingart (eds.), *Science, technology and the military*, pp. 307–322.

Solingen, Etel (ed.), *Scientists and the state: domestic structures and the international context*, Ann Arbor: University of Michigan Press, 1994.

Soll, Jacob, *The information master: Jean-Baptiste Colbert's secret state intelligence system*, Ann Arbor, University of Michigan Press, 2009.

Solomon, Julie Robin, *Objectivity in the making: Francis Bacon and the politics of inquiry*, Baltimore: Johns Hopkins University Press, 1998.

Solomon, Susan and Krementsov, Nickolai, 'Giving and taking across borders: the Rockefeller Foundation and Russia, 1919–28', *Minerva*, 39 (3), 2001, 265–98.

Spedding, James, *The letters and the life of Francis Bacon*, London: Longman, Green, Longman & Roberts, 1861.

Stapleton, Darwin H., 'Élève des Poudres: E.I. du Pont's multiple transfers of French technology' in Brenda Buchanan (ed.), *Gunpowder, explosives and the state: a technological history*, Aldershot: Ashgate, 2006, pp. 230–38.

Steen, Kathryn, 'Technical expertise and US mobilization 1917–18: high explosives and war gases' in MacLeod and Johnson (eds.), *Frontline and factory*, pp. 103–22.

Stewart, Irvin, *Organizing research for war*, Boston: Little Brown and Coy, 1948.

Storey, William E., 'Plants, power and development: founding the Imperial Department of Agriculture for the West Indies, 1880–1914' in Sheila Jasanoff (ed.), *States of knowledge: the co-production of science and social order*, New York: Routledge, 2004, pp. 109–30.

Storey, William Kelleher, *Science and power in colonial Mauritius*, Rochester: Rochester University Press, 1997.

Stoup, Alice, *A company of scientists: Botany, patronage, and community at the seventeenth-century Parisian Royal Academy of Sciences*, Berkeley: University of California Press, 1990.

'The political theory and practice of technology under Louis XIV' in Moran (ed.), *Patronage*, pp. 211–13.

Stranges, Anthony N., 'The US Bureau of Mines: synthetic fuel programme, 1920–1950s: German connections and American advances', *Annals of Science*, 54 (1), 1997, 29–68.

Sörlin, Sverker, 'Ordering the world for Europe: science as intelligence and information as seen from the northern periphery source', *Osiris*, 15, 2000, 51–69.

Szöllösi-Janze, Margit, 'Science and social space: transformations in the institutions of Wissenschaft from the Wilhelmite empire to the Weimar republic', *Minerva*, 43, 2005, 339–60.

Taton, René, 'Jean-Dominique Cassini' in Charles Gillispie (ed.), *Dictionary of scientific biography*, New York: Scribner 1970, vol. 3, pp. 100–6.

Terrall, Mary, 'The culture of science in Frederick the Great's Berlin', *History of Science*, 28 (4), 1990, 333–64.

Tilly, Charles (ed.), *The formation of national states in Western Europe*, Princeton: Princeton University Press, 1975.

'Reflections on the history of European state-making' in Charles Tilly (ed.), The formation of national states, pp. 1–83.

'Western state-making and theories of political transformation' in Charles Tilly (ed.), *The formation of national states*, pp. 601–38.

Tinkler, Hugh, 'The national state in Asia' in Tivey (ed.), *The nation-state*, pp. 104–21.

Tivey, Leonard, 'Introduction' in Tivey (ed.), *The nation-state*, pp. 1–12.

Tivey, Leonard (ed.), *The nation-state: The formation of modern politics*, Oxford: Martin Robertson, 1981, pp. 13–38.

Tuchman, Arleen, *Science, medicine and the state in Germany: the case of Baden 1815–1871*, Oxford: Oxford University Press, 1993.

Turner, Frank, 'Public science in Britain 1880–1919', *Isis*, 71 (4), 1980, 589–608.

Underwood, Matthew, 'Ordering knowledge, re-ordering empire: science and state formation in the English Atlantic world, 1650–1688', PhD dissertation, Harvard, 2010.

Varcoe, Ian, *Organising for science in Britain: a case study*, Oxford: Oxford University Press, 1974.

  'Scientists, government and organised research: the early history of the DSIR 1914–16', *Minerva*, 8, 1976, 192–217.

  'Co-operative Research Associations in British Industry, 1918–34', *Minerva*, 19 (3), 1981, 433–63.

Vessuri, Hebe, 'Science in Latin America' in John Krige and Dominique Pestre (ed.), *Science in the twentieth century*, pp. 839–87.

Vincent, Andrew, *Theories of the state*, Oxford: Basil Blackwell, 1987.

Vogel, Klaus A. and Rankin, Alisha, 'European expansion and self-definition' in Park and Daston (eds.), *The Cambridge history of science*, vol. 3, pp. 818–90.

Von Gizycki, Rainald, 'Centre and periphery in the international scientific community: Germany, France and Great Britain in the nineteenth century', *Minerva*, 11 (4), 1973, 474–94.

von Oertzed, Christine, 'Machineries of data power: manual versus mechanical census compilation in nineteenth century Europe', *Osiris*, 32, 2017, 129–15.

Vucinich, Alexander, *Science in Russian culture 1861–1917*, Stanford: Stanford University Press, 1970.

  Empire of knowledge: The Academy of Sciences of the USSR (1917–1970), Berkeley: University of California Press, 1984.

Wakefield, André, *The disordered police state: German cameralism as science and practice*, Chicago: University of Chicago Press 2009.

Walker, Mark, *German national socialism and the quest for nuclear power 1939–1949*, Cambridge: Cambridge University Press, 1989.

  Nazi science: Myth, truth, and the German atom bomb, New York: Plenum Press, 1995.

  'Twentieth century German science. Institutional innovation and adaptation' in Krige and Pestre (eds.), *Science in the twentieth century*, pp. 795–819.

  'Introduction' to special issue on 'Science in the Nazi regime: The Kaiser Wilhelm Society under Hitler', *Minerva*, 2006, 44, 241–50.

Wang, Jessica, 'Scientists and the problem of the public in Cold War America, 1945–1960', *Osiris*, 17, 2002, pp. 323–47.

  American science in an age of anxiety. Scientists, anticommunism, and the Cold War, Chapel Hill: University of New Carolina Press, 1999.

Watson, Mark F. and Noltie, Henry J., 'Career, collections, reports and publications of Dr. Francis Buchanan (later Hamilton), 1762–1829: natural history studies in Nepal, Burma (Myanmar), Bangladesh and India'. Part 1, *Annals of Science*, 73 (4), 2016, 392–424.

Weale, Adrian, *Science and the swastika*, London: Macmillan, 2001.

Webster, Charles, *The great instauration: science, medicine and reform, 1626–1660*, London: Duckworth, 1975.

Weindling, Paul, 'Medicine and modernisation: the social history of German health and medicine', *History of Science*, 24, 1986, pp. 277–301.

'The Rockefeller Foundation and German biomedical sciences, 1920–40: from educational philanthropy to international science policy' in Nicolaas A. Rupke (ed.), *Science, politics and the public good*, Houndmills: Palgrave, 1988, pp. 119–40.

Health, race and German politics between German Unification and Nazism 1870–1945, Cambridge: Cambridge University Press, 1989.

Werrett, Simon, 'The Schumacher affair: reconfiguring academic expertise across dynasties in eighteenth-century Russia', *Osiris*, 25 (1), 2010, 104–26.

Westfall, Richard S., 'Science and patronage: Galileo and the telescope', *Isis*, 76 (1), 1985, 11–30.

'Patronage and the publication of Galileo's *Dialogue*', *History and technology*, 4, 1987, 385–99.

Westman, Robert, 'The astronomer's role in the sixteenth century', *History of science*, 18, 1980, 105–47.

Westsick, Peter J., *The National Labs: Science in an American system, 1947–1974*, Cambridge, MA: Harvard University Press, 2003.

Whitney, Charles C., 'Merchants of light: science as colonization in the New Atlantis' in William E. Sessions (ed.), *Francis Bacon's legacy of texts*, New York: AMS Press, 1990, pp. 255–68.

Widmalm, Sven, 'Instituting science in Sweden' in Porter and Teich, *The scientific revolution*, pp. 240–62.

Widmalm, Sven, 'Accuracy, rhetoric, and technology: the Paris-Greenwich triangulation, 1784–88' in Frängsmyr, Heilbron and Rider (eds.), *The quantifying spirit*, pp. 179–206.

Wilkie, Tom, *British science and politics since 1945*, Oxford: Blackwell, 1991.

Williams, L. Pearce, 'Science, education and Napoleon I', *Isis*, 47 (4), 1956, 369–82.

Willmoth, Frances, *Sir Jonas Moore: Practical mathematics and Restoration science*, Woodbridge: Boydell Press, 1993.

Winch, Donald, 'The science of the legislator: the Enlightenment heritage' in Michael Lacey and Mary Furner (eds.), *The state and social investigation in Britain and the United States*, Cambridge: Cambridge University Press, 1993, pp. 63–91.

Windsor, Mary, 'Museums' in Bowler and Pickstone (eds.), *The Cambridge history of science*, vol. 6, pp. 60–75.

Wolfe, Audra J., *Competing with the Soviets: science, technology, and the state in Cold War America*, Baltimore: Johns Hopkins University Press, 2013.

Worboys, Michael, 'Science and the colonial empire, 1895–1940' in Kumar (ed.), *Science and empire*, pp. 13–27.

'Public and environmental health' in Bowler and Pickstone (eds.), *The Cambridge history of science*, vol. 6, pp. 141–64.

Wraight, A. Joseph and Roberts, Elliott B., *The Coast and Geodetic Survey 1807–1957*, Washington, DC: US Department of Commerce, 1957.

Yeo, Eileen Janes, 'Social surveys in the eighteenth and nineteenth centuries' in Theodore M. Porter, Dorothy Ross (eds.), *The Cambridge History of Science*:

vol. 7, *The modern social sciences*, Cambridge: Cambridge University Press, 2003, pp. 83–99.

Zachary, G. Pascal, *Endless frontier, Vannevar Bush: Engineer of the American century*, Cambridge, MA: MIT, 1999.

Zeldin, Theodore, 'Higher education in France, 1848–1940', *Journal of contemporary history*, 2 (3), 1967, 53–80.

Zwerling, Craig, 'The emergence of the École normale superieure as a centre of scientific education in nineteenth century' in Fox and Weisz (eds.), *The organization of science*, pp. 31–60.

Index

Note: Locators in italics denote illustrations.